GCSE in a week

Mathematics

Catherine Brown
and Lee Cope,
Abbey Tutorial College
Series Editor: Kevin Byrne

Where to find the information you need

SUCCESS OR YOUR MONEY BACK

Letts' market leading series GCSE in a Week gives you everything you need for exam success. We're so confident that they're the best revision books you can buy that if you don't make the grade we will give you your money back!

HERE'S HOW IT WORKS

Register the Letts GCSE in a Week guide you buy by writing to us within 28 days of purchase with the following information:

- Name
- Address
- Postcode
- Subject of GCSE in a Week book bought
- Probable tier you will enter

Please include your till receipt

To make a **claim**, compare your results to the grades below. If any of your grades qualify for a refund, make a claim by writing to us within 28 days of getting your results, enclosing a copy of your original exam slip. If you do not register, you won't be able to make a claim after you receive your results.

CLAIM IF...

You're a Higher Tier student and get a D grade or below.
You're an Intermediate Tier student and get an E grade or below.
You're a Foundation Tier student and get an F grade or below.
You're a Scottish Standard grade student taking Credit and General level exams, and get a grade 4 or below.
This offer is not open to Scottish Standard Grade students sitting Foundation level exams.

Registration and claim address:
Letts Success or Your Money Back Offer, Letts Educational, Aldine Place, London W12 8AW

TERMS AND CONDITIONS

1. Applies to the Letts GCSE in a Week series only
2. Registration of purchases must be received by Letts Educational within 28 days of the purchase date
3. Registration must be accompanied by a valid till receipt
4. All money back claims must be received by Letts Educational within 28 days of receiving exam results
5. All claims must be accompanied by a letter stating the claim and a copy of the relevant exam results slip
6. Claims will be invalid if they do not match with the original registered subjects
7. Letts Educational reserves the right to seek confirmation of the Tier of entry of the claimant
8. Responsibility cannot be accepted for lost, delayed or damaged applications, or applications received outside of the stated registration / claim timescales
9. Proof of posting will not be accepted as proof of delivery
10. Offer only available to GCSE students studying within the UK
11. SUCCESS OR YOUR MONEY BACK is promoted by Letts Educational, Aldine Place, London W12 8AW
12. Registration indicates a complete acceptance of these rules
13. Illegible entries will be disqualified
14. In all matters, the decision of Letts Educational will be final and no correspondence will be entered into

Letts Educational
The Chiswick Centre
414 Chiswick High Road
London W4 5TF
Tel: 020 8996 3333
Fax: 020 8742 8390
e-mail: mail@lettsed.co.uk
website: http://www.letts-education.com

First published 1998
New edition 2000
10 9 8 7 6

British Library Cataloguing in Publication Data
A CIP record for this book is available from the British Library.

ISBN 1 84085 3441

Design, artwork and production by Gregor Arthur
at Starfish Design for Print, London
Editorial by Tanya Solomons

Printed in Italy

Letts Educational is the trading name of Letts Educational Ltd, a division of Granada Learning Ltd. Part of Granada plc.

10 minutes

Test your knowledge

1 Write down

- **a** 2.376 to 2 d.p.
- **b** 27.77 777 to 3 d.p.
- **c** 4.761 11 to 3 s.f.
- **d** 236 000 to 2 s.f.

2

- **a** What is 32% of £72?
- **b** Mark scored 73 marks out of 150 on the Biology test. Give his percentage score to 1 d.p.
- **c** Lala weighs 3.6 kg at birth. A year later, Lala weighs 6.4 kg. Calculate the percentage increase in Lala's weight, giving your answer correct to 3 s.f.
- **d** Originally a calculator cost £30. In a sale, the price is reduced by 15%. Calculate the sale price of the calculator.
- **e** Kieron's car insurance premium is £834.99 in 1997, which is an increase of 30% on the figure in 1996. Calculate how much Kieron paid in 1996.

Answers

1 a) 2.38. b) 27.778. c) 4.76. d) 240 000.
2 a) £23.04. b) 48.7%. c) 77.8%.
d) £25.50. e) £642.30.

✔ If you got them all right, skip to page 5

Number work 1

Improve your knowledge

30 minutes

1 Decimal places/significant figures

Many exam questions ask you to express your answer to a certain number of decimal places or significant figures – if you don't do this right you won't get the marks!

Decimal places

We count the number of figures after the decimal point, starting next to the decimal point.

Example Write 3.376 134 to 2 decimal places.
We count to 2 places after the point and underline the next figure 3.37<u>6</u> 134.
If the number underlined is **5 or bigger**, we round **up**.
If the number underlined is **4 or smaller**, we round **down**.

In this example, the number underlined is 6, so we round **up** to give **3.38** (2 d.p.).

Significant figures

We start to count from the **first non-zero number** from the left and use the underlining rule from above.

Example Write 47.371 correct to 3 significant figures.
Counting 47.3<u>7</u>1
7 is bigger than 5 – round up.
So answer is 47.4 (3 s.f.).

Example Write 0.000 389 21 correct to 4 significant figures.
Counting (ignore the zeros at the front) 0.000 389 2<u>1</u>
1 is smaller than 4 – round down.
So the answer is 0.000 389 2 (4 s.f.).

Example Write 174 600 to 2 significant figures.
Counting 17<u>4</u> 600
4 is smaller than 5 – round down.
So the answer is 170 000 (2 s.f.).

Answer can't be 17 because the original number is nowhere near 17.

For a big number, change the digits we don't want to zeros.

2 Percentages

The word 'per cent' means 'out of 100' – 100% is the whole, original amount. There are 4 types of percentage questions.

Type 1 Standard

% means 'divide by 100' and 'of' means 'times'.

Example Calculate 30% of £500

$$30\% \text{ of } £500 \rightarrow \frac{30}{100} \times £500 = £150.$$

Example 18 students out of a Sixth Form of 75 study dance. What percentage is this?

$$\frac{18}{75} \times 100\% = 24\%$$

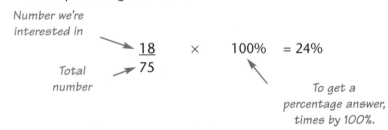

Number we're interested in (18)

Total number (75)

To get a percentage answer, times by 100%.

Type 2 Percentage change

You can be asked to find a **percentage increase** or **decrease**, or **profit** or **loss.** You need to **learn** the following formula.

$$\text{Percentage change} = \frac{\text{change}}{\text{original}} \times 100\%$$

Original always goes on the BOTTOM.

Example Lee bought a car for £3850. Ten months later, he sold the car for £3200. Calculate his percentage loss, giving your answer correct to 2 decimal places.

Change = 3850 – 3200 = £650 loss.

So % loss = $\frac{650}{3850} \times 100\% = 16.883\,116...\ \%$.

To 2 decimal places 16.88<u>3</u> 116 so 16.88%.

Use the same rule for finding a percentage <u>increase</u> (see 2c in 'Test your knowledge').

Type 3 Increase/decrease by a percentage

To do these, you need to go through two steps.

Step 1 Work out the percentage increase (or decrease).
Step 2 Add (or take away) from the original amount.

Example Sheila puts £2000 into a savings bond. Each year, the value of her investment increases by 8%.

a Calculate the value of her investment at the end of the first year.
The interest earned is 8% of £2000.

$$8\% \text{ of } 2000 = \frac{8}{100} \times £2000 = £160.$$

So the value at the end of the first year is £2000 + £160 = **£2160**.

b Calculate the value of her investment at the end of the second year. At the beginning of the second year, Sheila has £2160.

The interest earned is 8% of £2160.

$$8\% \text{ of } £2160 = \frac{8}{100} \times £2160 = £172.8.$$

So at the end of the second year, her investment is worth
£2160 + £172.80 = **£2332.80**.

Type 4 Finding the original amount before a percentage change

In these questions you are asked to find the **original value** before something has been increased or decreased by a percentage.

The original amount is 100%.

Example A surfboard has been reduced by 20% to £150. What is the original price?

The surfboard is reduced by 20%, so we have 100% – 20% = 80% left which is £150.

The original price is 100% which we do not know as a value in pounds.

Step 1 Draw a 2 × 2 table.
Step 2 Fill it in.
Step 3 Draw a cross.
Step 4 Do the calculation.

	Price	Percentage
Know	150	80
Want	?	100

$$\text{Answer} = \frac{150 \times 100}{80}$$
$$= £187.5$$
$$= £187.50.$$

Same as 50p

Times the joined numbers and divide by the other one.

Now learn how to use your knowledge

Number work 1

10 minutes

Use your knowledge

1 Barry has £125 000 in his bank account. **Hints 1/2**

a He decides to spend 17% of his money on a BMW car! How much does he pay for the car?

b He decides to buy a house which has been reduced from £70 000 to £55 000. Calculate the percentage reduction in the house-price, giving your answer to 3 significant figures. **Hints 3/4**

c After buying his house and car, what percentage of his original money has Barry got left? **Hints 5/6/7**

2 *a* Niall's electricity bill was £76.68 including VAT at 8%. What was his bill before VAT was added? **Hints 8/9/10**

b Niall's flatmate Martin pays 40% of the bill. How much does Martin pay (inclusive of VAT)? Give your answer correct to 2 decimal places. **Hint 11**

c Niall's friend Suzie has a bill 15% larger than Niall's. How much is Suzie's bill (inclusive of VAT)? (Give your answer correct to nearest penny). **Hints 12/13/14**

✓ Hints and answers follow

Number work 1

Hints

1 Work out 17% of his money.

2 What does per cent mean?

3 This is a type 2 percentage question.

4 Remember to divide by the original price.

5 Work out how much money Barry has spent on both items.

6 Work out how much money is left.

7 Find the percentage! What do you do when you're looking for a percentage?

8 Draw a table and fill it in.

9 The £76.68 is 108%.

10 Draw a cross.

11 Type 1 percentage question.

12 Type 3 percentage question.

13 Work out 15% of Niall's bill.

14 Add this to the bill.

Answers

1 a) $\frac{17}{100}$ × 125 000 = £21 250. b) Change = 70 000 – 55 000 = 15 000. Answer = 15 000/70 000 × 100% = 21.4%. c) Total spent = 21 250 + 55 000 = £76 250. Amount remaining = 125 000 – 76 250 = 48 750. Percentage remaining = 48 750/125 000 × 100% = 39%.
2 a) $\frac{£76.68 \times 100}{108}$ = £71. b) $\frac{40}{100}$ × £76.68 = £30.67. c) $\frac{15}{100}$ × £76.68 = £11.50. Total = £76.68 + £11.50 = £88.18.

10 minutes

Test your knowledge

1 **a** £1 is worth 3.05 German Deutschmarks. What is the value of £37.50 in German DMs?

b There are 1760 yards in a mile and 1.6 km in a mile. How many yards are there in 10 km?

2 Give the following numbers in standard form.

a 3 310 000

b 760 000

c 136.71

d $3 \times 10^4 \times 8 \times 10^5$

e 0.000 361

f 0.0014

g 40 000

h 0.000 006 67

3 Using the method of trial and improvement, solve this equation

$$x^3 + x^2 = 50$$

giving your answer correct to 1 d.p. Use starting values of x = 0 and x = 5.

Answers

1 a) 114.38 DMs. b) 11 000 yards.
2 a) 3.31×10^6. b) 7.6×10^5.
c) 1.3671×10^2. d) 2.4×10^{10}. e) 3.61×10^{-4}.
f) 1.4×10^{-3}. g) 4×10^4. h) 6.67×10^{-6}.
3 3.4.

If you got them all right, skip to page 11

Number work 2

Improve your knowledge

30 minutes

1 Conversions

To do conversions, use a 2 × 2 table as in the type 4 percentage example on page 4

Example £1 is worth 1.65 American dollars.
What is the value of a $100 note in pounds?

Step 1 Draw a table.

	£	$
Know	1	1.65
Want	?	100

Step 2 Draw a cross.

Step 3 Multiply the joined numbers, divide by the other one.

$\frac{1 \times 100}{1.65} = £60.61$ (nearest penny).

Example There are 3 feet in 1 yard, 12 inches in 1 foot and 39.37 inches in 1 metre. How many yards are there in 8 metres? (Give your answer to 3 significant figures).

We know 3 × 12 = 36 inches in one yard.
8 metres = 8 × 39.37 = 314.96 inches =?? yards.

Draw a table:

	Inches	Yards
Know	36	1
Want	314.96	?

Answer = $\frac{314.96 \times 1}{36}$ = 8.74<u>8</u>88

= 8.75 yards (3 s.f.)

2 Standard form

Standard form is just a quicker way to write very big numbers or very small numbers.

For example there are approximately 54 000 000 people in the UK. It's boring to write out all those zeros (and we might miss one out), so instead, we write it in standard form.

$$5.4 \times 10^7$$

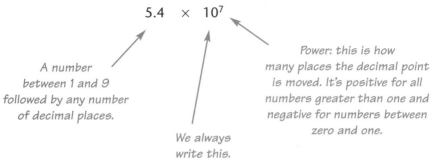

A number between 1 and 9 followed by any number of decimal places.

We always write this.

Power: this is how many places the decimal point is moved. It's positive for all numbers greater than one and negative for numbers between zero and one.

You can put this in your calculator by pressing: [5] [•] [4] [EXP] [7] [=]

Some calculators use [EE] instead of [EXP]

Example Write 3 610 000 in standard form.

Step 1 Put a decimal point to the right of the first (non-zero) number.
Step 2 Count how many spaces the decimal point has to move to get back to where it was before.

We start with 3 610 000.

Put the point in after the 3: 3.610000 we've made 6 jumps.

So the answer is 3.61×10^6 (positive power since it is a big number).

Example Write 0.000 000 26 in standard form.

Put the point after the 2: 00000002.6 we've made 7 jumps.

So the answer is 2.6×10^{-7} (negative power since it is a small number).

Example Calculate $2.8 \times 10^5 \times 4 \times 10^{-3}$, expressing your answer in standard form.

Press the [±] button after [EXP] button and before the power to get negative powers in your calculator.

Type in: [2] [•] [8] [EXP] [5] [×] [4] [EXP] [±] [3] [=]

You get 1120.
But we must express this in standard form.
We have to move the decimal point 3 places – so the answer is 1.12×10^3.

3 Trial and Improvement

This is a way of solving equations that are too hard to do by algebra. All it involves is putting numbers into an equation, setting this out in a table and looking at the result.

Example Use a method of trial and improvement to solve $x^3 + x = 60$ correct to 1 decimal place, starting with $x = 4$ and $x = 3$.

	Value	Calculation and answer	Bigger or smaller than 60
Start by using given numbers	4	$4^3 + 4 = 68$	Too big – try smaller!
	3	$3^3 + 3 = 30$	Too small – try bigger!
Number in between 3 and 4	3.5	$3.5^3 + 3.5 = 46.375$	Too small – try bigger!
Number in between 3.5 and 4	3.8	$3.8^3 + 3.8 = 58.672$	Too small – try bigger!
Number in between 3.8 and 4	3.9	$3.9^3 + 3.9 = 63.219$	Too big – try smaller!
Number in between 3.8 and 3.9	3.85	$3.85^3 + 3.85 = 60.9166$	Too big

The answer lies between 3.8 and 3.85. All of these values round to **3.8** to 1 d.p. So the answer is 3.8.

Always show all the table working in the exam.

✔ Now learn how to use your knowledge

Number work 2

10 minutes

Use your knowledge

1 Mr Fat Cat has 2.6×10^6 shares in the insurance company 'Quotes R Us'.

a Mr Fat Cat wants to sell some shares to give him £100,000. 100 shares sell for £650. How many shares must he sell? Give your answer to 4 significant figures. Hints 1/2

b How many shares does Mr Fat Cat have left? Give your answer in standard form. Hints 3/4/5

2 Mohsin set his friends a puzzle. He said:

'I have thought of a number. If I cube the number, add 2, then subtract the number I first thought of, I get 1.2146×10^4. What is the number I first thought of?'

To answer this question, Kulvinder wrote down the equation

$$x^3 + 2 - x = 1.2146 \times 10^4.$$

He decided to solve it by trial and improvement.

a First of all, he converted 1.2146×10^4 to a number not in standard form. What answer did he get? Hints 6/7

b He tried $x = 20$ and $x = 30$ first. Work out what answers he got. Hint 8

c Use trial and improvement to find the number Mohsin first thought of. Hints 9/10/11

✓ Hints and answers follow

Number work 2

Hints

1 Draw a table and a cross.

2 Times the two joined numbers, divide by the other one.

3 Subtract what he sold from what he started with.

4 Use the [EE] or [EXP] button to enter numbers in standard form on your calculator.

5 Remember – jump the decimal point.

6 How many places do you have to jump the decimal point?

7 Which way do you have to jump the decimal point?

8 Put x = 20, then x = 30 in Kulvinder's equation.

9 20 is too small and 30 too big. What should you do?

10 Try a number in between 20 and 30.

11 Keep on going until you get the answer.

Answers

1 a) $\frac{£100\,000 \times 100}{£650}$ = 15 384.62 = 15 380 to 4 s.f. b) $2.6 \times 10^6 - 15\,380 = 2\,584\,620 = 2.584\,62 \times 10^6$.
2 a) 12 146. b) x = 20: answer = 7982 (too small). x = 30: answer = 26 972 (too big).
c) x = 25: answer = 15 602 (too big). x = 22: answer = 10 628 (too small).
x = 23: answer = 12 146, so answer = 23.

Angles

10 minutes

Test your knowledge

1 Find the angles marked with letters.

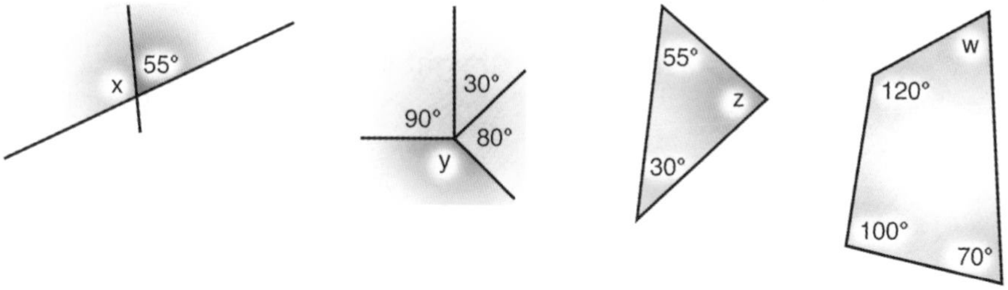

2 Find the angles marked with letters.

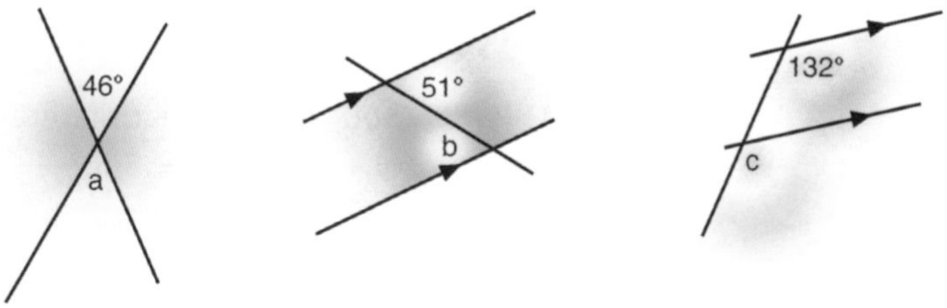

3 I want to make a tessellation using hexagons like this one. All its angles are the same.

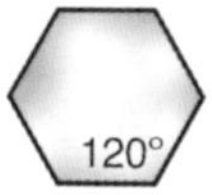

Explain why the tessellation will work.

Answers

1 x = 125°, y = 160°, z = 95°, w = 70°.
2 a = 46°, b = 51°, c = 132°. **3** When we make the tessellation, we will be joining together three 120° angles, which add up to 360°. Since they add up to 360°, the tessellation works.

If you got them all right, skip to page 17

Angles

Improve your knowledge

30 minutes

1 Important angle facts – part 1

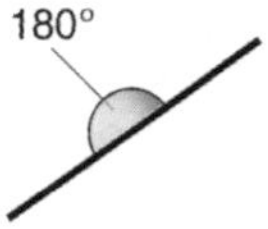

180° in a straight line.

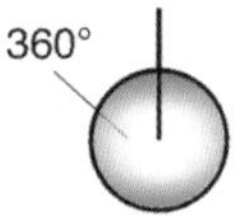

360° in a full circle.

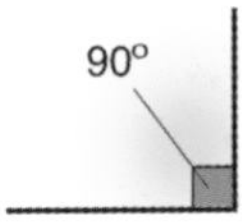

90° in a right-angle.

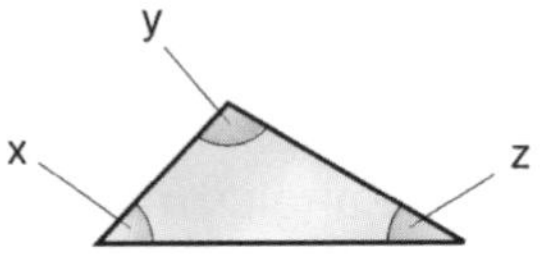

80° in a triangle.
x + y + z = 180°

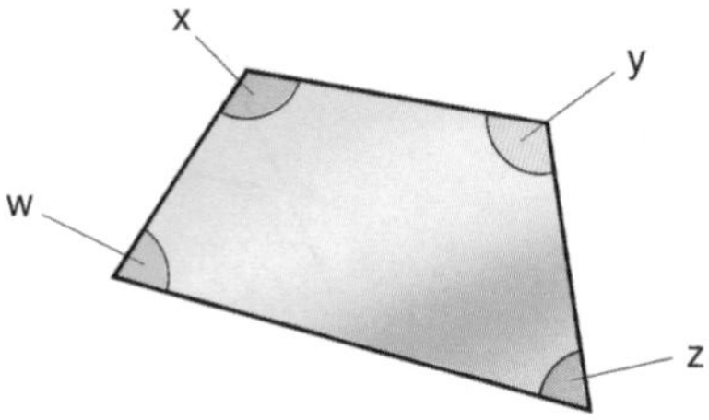

360° in any quadrilateral.
w + x + y + z = 360°

Using these facts

Must be **63°**,
since 63° = 360° – 120° – 92° – 85°.

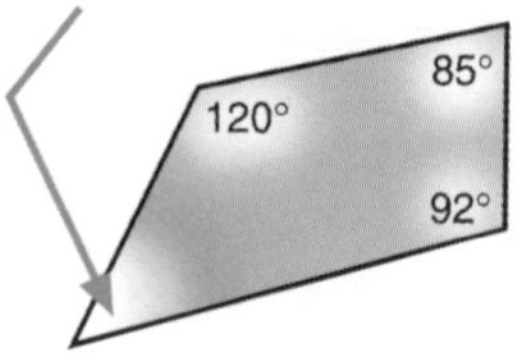

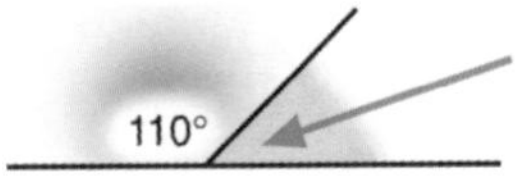

Must be **70°**,
since 70° = 180° – 110°.

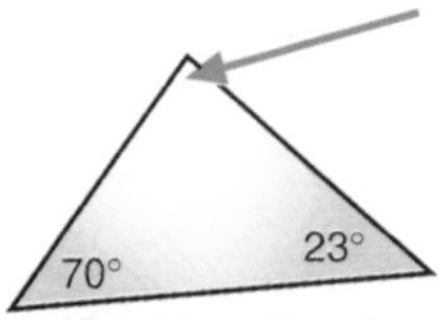

Must be **87°**,
since 87° = 180° – 70° – 23°.

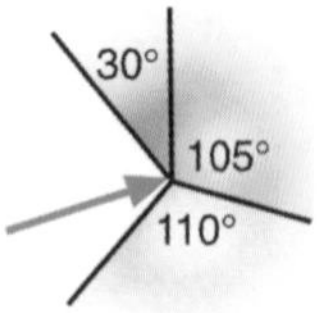

Must be **115°**,
since 115° = 360° – 30° – 105° – 110°.

2 *Important angle facts – part 2*

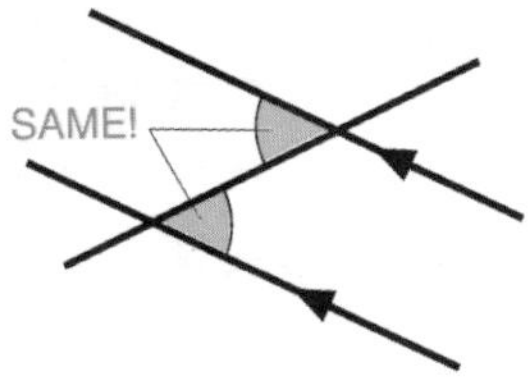

These are **Z angles** – they are equal.
These lines have to be parallel or it won't work!

Those arrows mean lines are PARALLEL.

NB. *The Z can be the right way round or backwards.*

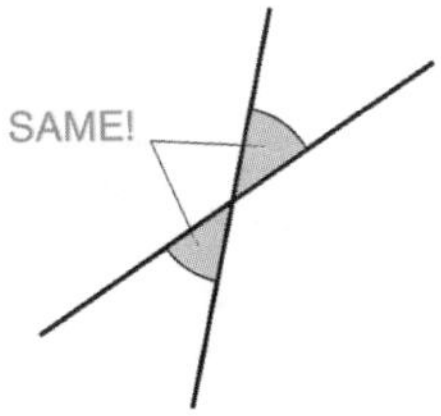

These are **Opposite angles** – they are equal.

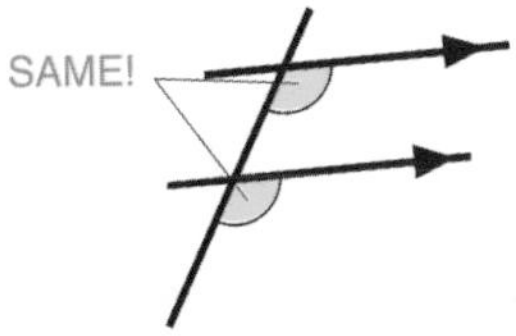

These are **F angles** – they are equal.
The lines have to be parallel here too!

Using these facts

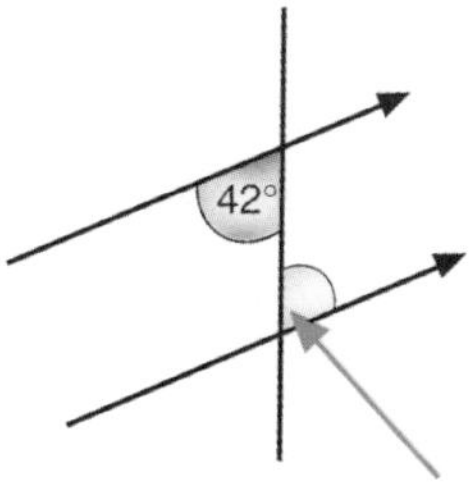

Must be **42°**,
since we have Z angles.

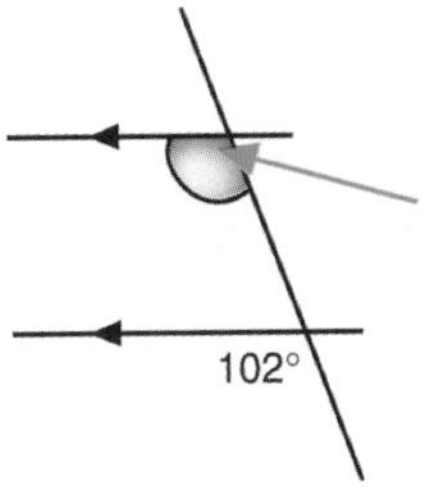

Must be **102°**,
since we have
F angles.

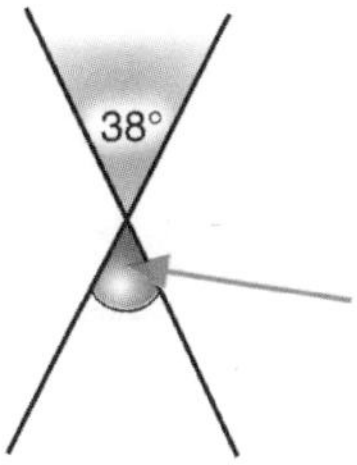

Must be **38°**, since we have opposite angles.

We can use more than one rule to answer a question.

e.g. Find the angles x, y and z shown.

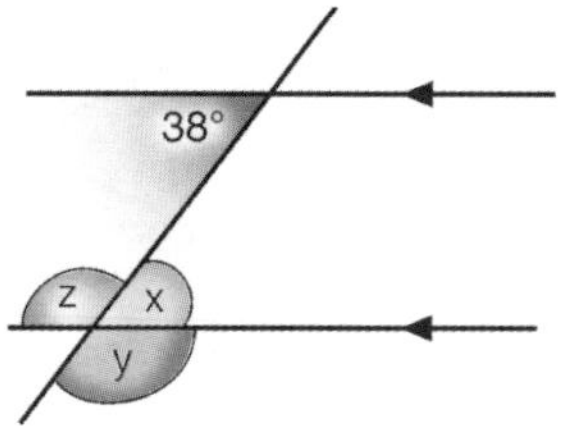

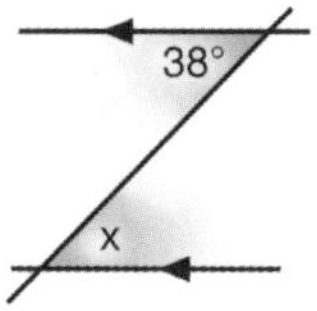

Look to see Z angles or F angles.

Because we've found a Z angle, we get x = **38°**.

Now we must try to find the others.
Remember we can use the value we know for x!

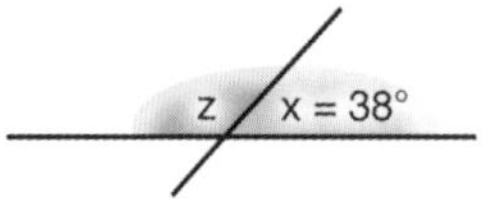

z and x make up a straight line.
So they must add up to 180°.
So z = 180° – 38° = **142°**.

Now we need y. We can use the values for x and z.

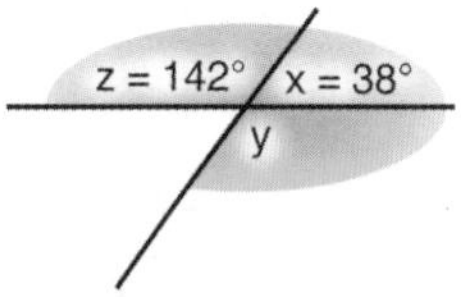

y is opposite z. So z and y are equal.
So y = **142°**.

3 *Tessellations*

One final use for angles is to see if **tessellations** work.

A tessellation will only work if the interior angles you are trying to fit together add up to 360°.

Remember ***tessellation*** *means fitting shapes together without gaps.*

e.g. If you try to fit together four squares, you are combining four 90° angles. This adds up to 360°, so the tessellation works.

Now learn how to use your knowledge

Angles

20 minutes

Use your knowledge

1 This is a regular pentagon (**regular** means that all the angles are the same).

Point P is at the middle of it.

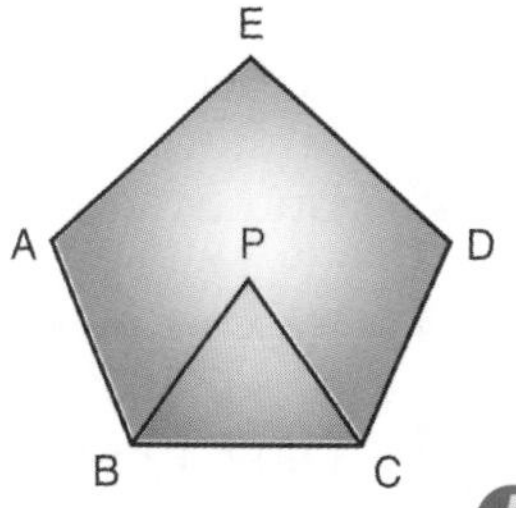

a Explain why triangle PBC is isosceles. — Hints 1/2

b What is angle BPC? — Hints 3/4

c What is angle PBC? — Hints 5/6

d What is the size of an inside angle of the pentagon? — Hint 7

e Is it possible to make a tessellation from these pentagons? — Hints 8/9

2 The diagram shows a shape made from putting a triangle (CDE) on top of a rectangle (ABCE). The corners of the rectangle and angles DEB and DCA are right-angles, and angle EBC is 55°.

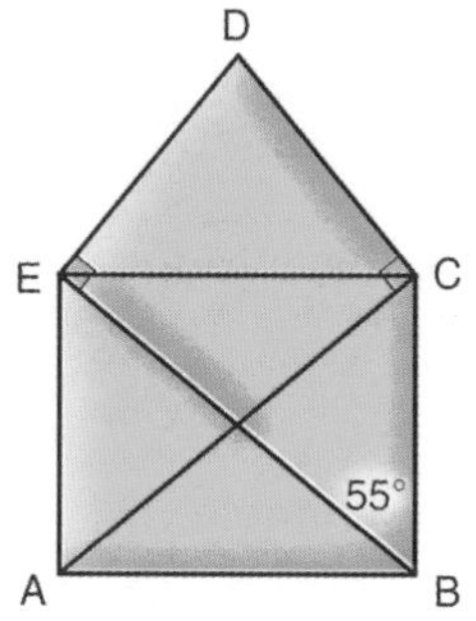

Find

a angle ABE — Hints 10/11

b angle CEB — Hint 12

c angle DEC — Hint 13

d angle DCE — Hints 14/15

e angle CDE. — Hint 16

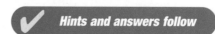

Hints and answers follow

Angles

Hints

1 Remember an isosceles triangle has two sides the same.

2 Find two sides that are the same.

3 How many of this angle do you need to make 360°?

4 Join P to the other corners of the pentagon.
The angles at the middle are all the same as each other.

5 What do the angles of a triangle add up to?

6 What other angle is the same as angle PBC?

7 You can get one of these angles by doubling angle PBC.

8 What needs to be true if the tessallation is to work?

9 Try adding up the angles for different numbers of pentagons.

10 Remember the letter in the middle tells you where the angle is.

11 What must this angle and the one you have been given add up to?

12 Look for Z angles.

13 What must angles DEC and CEB add up to?

14 This is the same as another angle you have found – which one is it?

15 Use symmetry.

16 You know the other two angles of the triangle.

Answers

1 a) Because PB and PC are the same length (they must be the same length because P is the middle). b) Five lots of this angle must come to 360°, so the angle is equal to 360/5 = 72°. c) PBC + PCB + 72 = 180°, so PBC + PCB = 180 − 72 = 108°. But PBC and PCB are the same, so 2 × PBC = 108°, so PBC = 108/2 = 54°. d) Two times angle PBC = 108°. e) No, because three add up to 3 × 108° = 324°, and four add up to 4 × 108° = 432°, so we can't get it to add up to 360°. **2** a) ABE + 55° = 90°, so ABE = 35°. b) Z angles – angle CEB and angle EBA. So they are equal, so CEB = 35°. c) DEC + 35° = 90°, so DEC = 55°. d) Because the whole thing is symmetrical, this is the same as DCE, so it is 55°. e) 55° + 55° + CDE = 180°, so CDE = 70°.

Algebra 1

10 minutes

Test your knowledge

1 Simplify $4 - 3x + 5 + 4x$.

2 Find x if $19 - 3x = 13$.

3 Find x if $\frac{x^2}{3} = 27$.

4 Find x if $7 - 6x = 8 + 3x$.

5 Make y the subject of the formula $1 - 2y = 3x$.

6 Make x the subject in the following equations:

a $d = ax + b$.

b $v^2 = u^2 + 2ax$.

Answers

1 $9 + x$. **2** $x = 2$. **3** $x = 9$. **4** $x = -1/9$.
5 $y = \frac{1}{2} - \frac{3x}{2}$ or $y = \frac{1-3x}{2}$. **6** a) $x = \frac{d}{a} - \frac{b}{a}$ or $x = \frac{d-b}{a}$. b) $x = \frac{v^2}{2a} - \frac{u^2}{2a}$ or $x = \frac{v^2-u^2}{2a}$.

✔ If you got them all right, skip to page 23

Algebra 1

Improve your knowledge

30 minutes

Algebra basics

Before we can do any algebra, we need to learn how to work with letters. The first thing to remember is that we can't add or subtract letters and numbers and get a single answer – it's like adding an apple to an orange and getting 2 apples as an answer! We can also only add or subtract letters of the same kind – we can't combine x and y, or even x and x^2.

1 We can simplify things like $5x - 2x$. This gives $3x$ (think of it as 'five xylophones' – 'two xylophones' = 'three xylophones').

Example Simplify $3x - 2 + 4x - 5 + 6x^2$.

Step 1 Collect together like terms.

This means that you put all the x bits together, all the x^2 bits together, and all the numbers together.

$3x + 4x + 6x^2 - 2 - 5$

Step 2 $= 7x + 6x^2 - 7$.

All we are doing in algebra is just using letters to stand for numbers that we don't know – using x is just like using ? or □

Algebra questions will ask you to do one of two things:

- find out a value for a letter,
- make a particular letter the **subject** of a formula.

The SUBJECT of a formula is the letter on its own on one side of the equals sign.

These really aren't too different to each other! In both of them, we will be trying to move things around in the equation to get one letter on its own.

We can often answer the easiest sort of question by thinking about it like this:

Example 1 The equation $2 + x = 10$ can be solved by saying to ourselves 'what do I have to add to 2 to get 10?' – which hopefully gives the answer 8.

Example 2 The equation $3x = 27$ can be solved by asking ourselves 'what do I have to times by 3 to get 27?' – which leads to the answer 9.

If it is more complicated, you will probably need to use the rules. These are the same for both types of question.

Remember – we are trying to get x on its own.

2 What are the rules?

We will illustrate the rules by looking at some examples.

Example 1	$2x + 3 = 7$	First we need to get rid of that '+3'
	$2x = 7 - 3$	To get rid of a plus, **take it to the other side and it becomes minus**.
So	$2x = 4$	Now we need to get rid of the '2' in front of the x.
	$x = \frac{4}{2}$	To get rid of a number in front, **take it to the other side and divide**.
So	$x = 2$	
Example 2	$\frac{x}{3} - 4 = 1$	First we need to get rid of that '–4'
	$\frac{x}{3} = 1 + 4$	To get rid of a minus, **take it to the other side and it becomes plus**.
So	$\frac{x}{3} = 5$	Now we need to get rid of the '3' underneath x.
	$x = 5 \times 3$	To get rid of a number underneath, **take it to the other side and times**.
So	$x = 15$	
Example 3	$2 - x = 4$	First get rid of the '2'. It is really a '+ 2' but we just don't show the '+' because the 2 is at the start.
	$-x = 4 - 2$	It's a **plus**, so it becomes a **minus** on the other side.
	$-x = 2$	
We must always sort out the '+' or '–' first.	$x = -2$	We have a minus sign in front of the x! To get rid of this, **change the sign on the other side**.

Some questions may try to make things harder for you by putting things with x in on both sides. **Don't panic!** All you have to do is apply the rules to put both bits with x in on the same side. Let's see ...

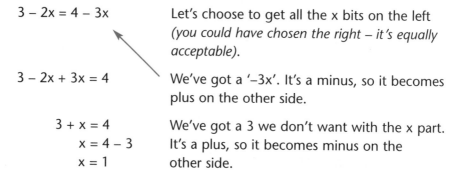

$3 - 2x = 4 - 3x$ Let's choose to get all the x bits on the left *(you could have chosen the right – it's equally acceptable).*

$3 - 2x + 3x = 4$ We've got a '$-3x$'. It's a minus, so it becomes plus on the other side.

$3 + x = 4$

$x = 4 - 3$

$x = 1$

We've got a 3 we don't want with the x part. It's a plus, so it becomes minus on the other side.

The other thing we need to use the rules for is **rearranging formulas**. It's the same idea – we are trying to get one letter (called the subject of the formula) on its own. The only difference is that the things we are moving around may be letters as well as numbers.

Example 4

Make x the subject of the formula $y = 2x + 1$. We want x on its own.

$y - 1 = 2x$ First we must get rid of the 1; plus becomes minus.

$\frac{y}{2} - \frac{1}{2} = x$ Now we need to get rid of the 2; divide by it on the other side.

x is on its own. We've finished!

Example 5

Make t the subject of the formula $v = u + at$. We want t on its own.

$v - u = at$ First we must get rid of the 'u'; plus becomes minus.

$\frac{v}{a} - \frac{u}{a} = t$ Now we must lose the 'a' in front; take across and divide.

✔ *Now learn how to use your knowledge*

Algebra 1

10 minutes

Use your knowledge

1 Navneet and Nadia are playing a number game. Navneet says to Nadia,

'Think of a number, double it, add 3.'

Navneet then asks Nadia what she has got, and works out the number that Nadia was originally thinking of.

a Call Nadia's original number 'x'. Write down a formula for the answer she gets in terms of x. Hints 1/2

b If Nadia's final answer is 11, what number did she think of? Hints 3/4/5/6

2 Ruby's mum always thinks in inches, but Ruby is only used to measuring in centimetres. Ruby knows that she can convert inches to centimetres by using the formula

$$c = \frac{100\,i}{39}$$

What formula should Ruby's mum use to convert centimetres to inches? Hints 7/8/9

3 Victoria is studying physics. She has learnt the formula

$$s = ut + \tfrac{1}{2}at^2$$

where s stands for distance, u stands for initial speed, a stands for acceleration and t stands for time.

In a question, she is given values for s, t and a, and she needs to find u. Work out a formula that will let Victoria find u. Hints 10/11/12/13

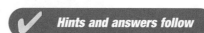

Hints and answers follow

Algebra 1

Hints

1 What is the first thing that is done to x?

2 'Double it' means 'times by …'

3 Use your answer to part a) to make an equation.

4 Get rid of the 3 first.

5 Plus on one side becomes what on the other?

6 How do we get rid of a number in front of the x?

7 Rearrange the formula to make i the subject.

8 How do we get rid of a number underneath?

9 How do we get rid of a number in front?

10 You need to make u the subject of the formula.

11 Get rid of the $\frac{1}{2}at^2$ first.

12 Plus on one side becomes …

13 How do we get rid of the t to leave u on its own?

Answers

1 a) Answer = 2x + 3. b) We have 11 = 2x + 3, so 11 − 3 = 2x, so 8 = 2x, so 8/2 = x, so 4 = x.

2 $c = \frac{100\,i}{39}$, so $39c = 100\,i$, so $\frac{39c}{100} = i$.

3 $s = ut + \frac{1}{2}at^2$, so $s - \frac{1}{2}at^2 = ut$, so $\frac{s}{t} - \left(\frac{1}{2}\frac{at^2}{t}\right) = u$, and $\frac{s}{t} - \frac{1}{2}at = u$.

10 minutes

Test your knowledge

1

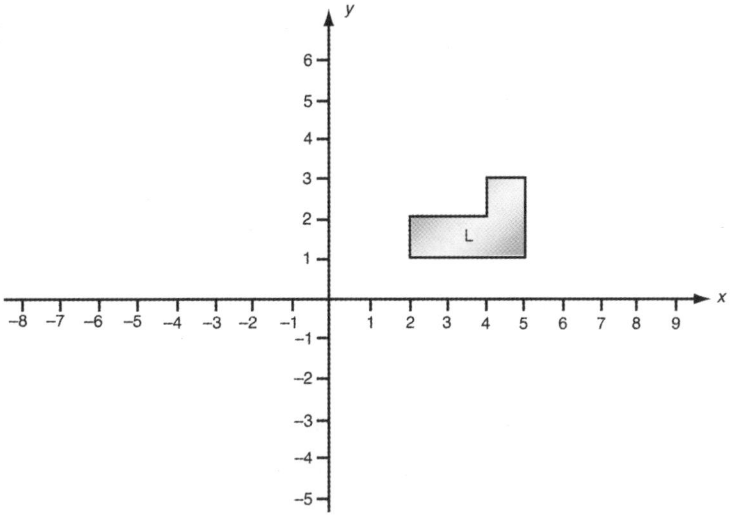

Draw the image of the shape L under:

a rotation 90° anticlockwise about (0,0).

b Translation $\begin{pmatrix} -2 \\ -6 \end{pmatrix}$.

c Reflection in the x-axis.

d Reflection in the line y = x.

e Enlargement scale factor 2, centre (1,1).

Answer

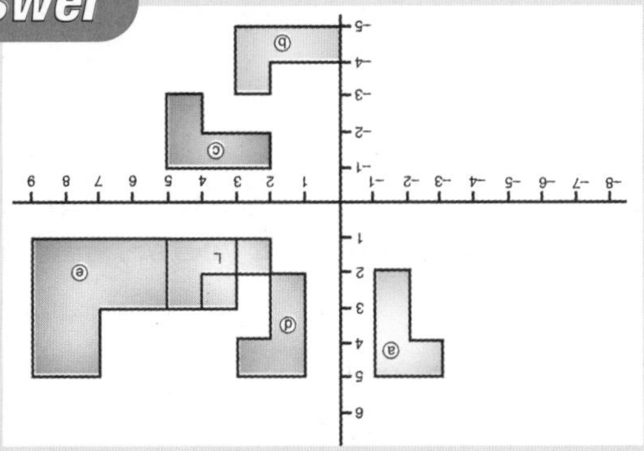

If you got them all right, skip to page 29

Transformations

30 minutes

Improve your knowledge

1 Transformations

There are four types of transformation you need to know.

1 Rotation

To describe a rotation, you need three things: **angle; direction; centre of rotation.**

Example 1

Describe the transformations

a A to B.
Rotation 90° clockwise about centre (0,0).

b C to B.
Rotation 180° (anti)clockwise about centre (0,0).

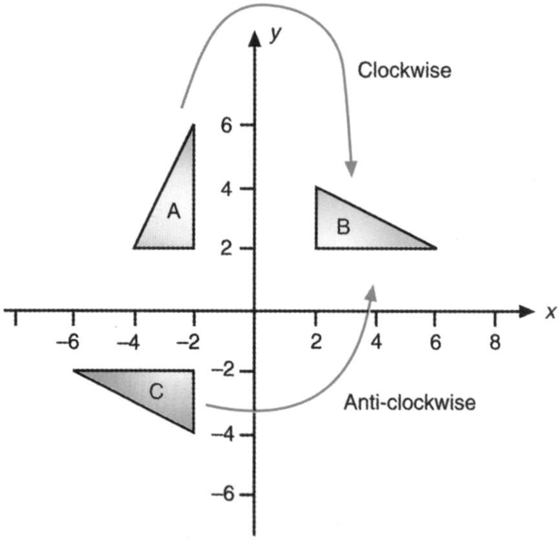

You can do rotations by using tracing paper – we'll do A to B:

Step 1 Trace out shape A and the centre of rotation (0,0).

Step 2 Put your pen on the centre of rotation and turn your tracing paper a quarter turn to the right.

Step 3 Rub down hard on your tracing paper and you should find shape B.

2 Translation

This just means moving a shape from one place to another. To do a translation, you need to know a **translation vector**, which is:

$\begin{pmatrix} x \\ y \end{pmatrix}$ x: distance moved left or right. y: distance moved up or down.

Positive numbers mean right or up. Negative numbers mean left or down.

Example 2

Describe the transformations

a D to E.
We've gone right 2 and up 4. So it's a translation $\binom{2}{4}$.

b E to F.
We've gone right 3 and down 5. So it's a translation $\binom{3}{-5}$.

c F to D. We've gone left 5 and up 1. So it's a translation $\binom{-5}{1}$.

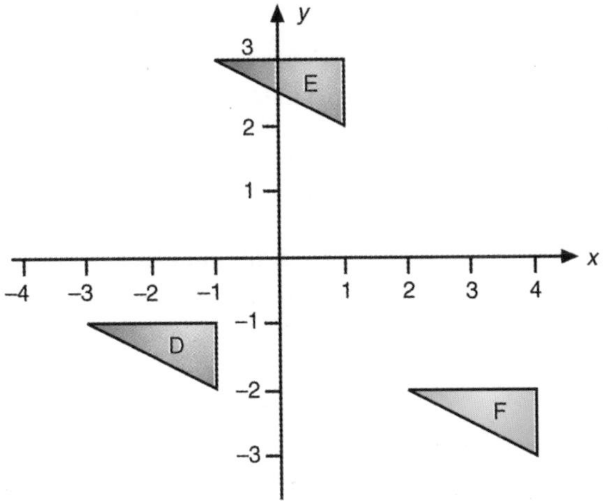

3 Reflection

A reflection is a shape produced by a mirror image. The **mirror line** is called the **axis of symmetry**.

To describe a reflection you must state the mirror line.

Example 3

Describe the transformations

a G to H.
Reflection in x-axis.

b G to J.
Reflection in y-axis.

c D to G.
Reflection in y = –x.

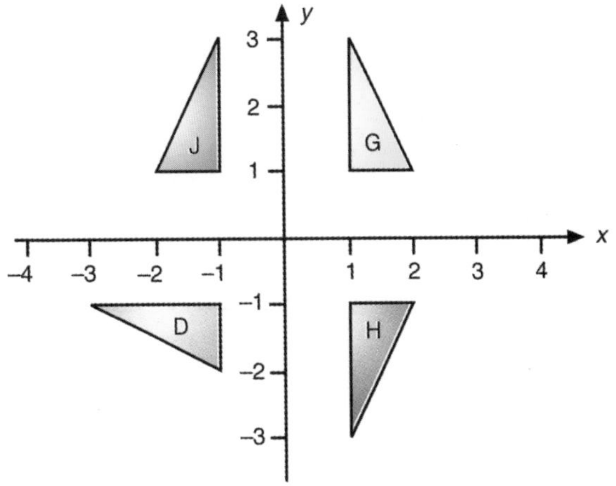

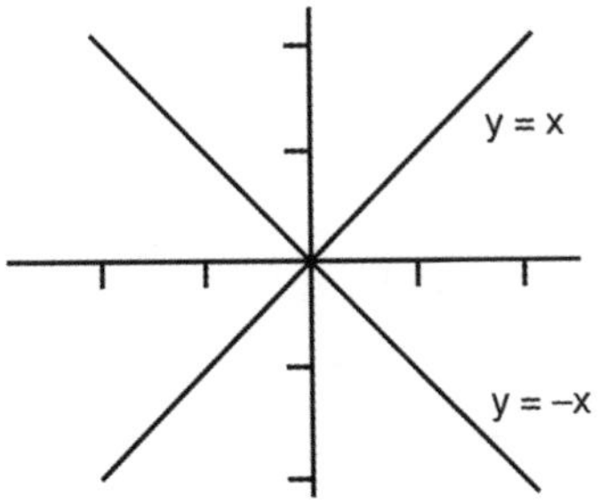

Remember Learn these special lines for reflections.

4 Enlargement

An enlargement is where the image produced is either a bigger version or a smaller version of the original.

To describe an enlargement you must state the **scale factor** and **centre of enlargement**.

The scale factor is how many times bigger the new shape is. You can find it by using the formula:

scale factor = new length/old length

The centre of enlargement is found by:

1 Joining the corresponding corners of the old shape and the new shape.

2 Extending the lines until they cross.

The centre of enlargement is the point where the lines cross.

Example 4

Describe the transformation K to L

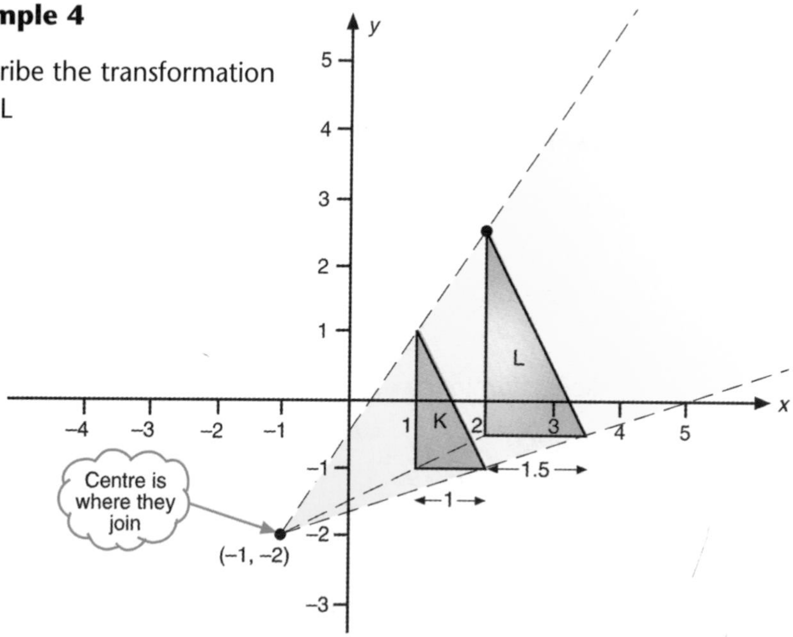

New base length = 1.5   Old base length = 1.
So scale factor = new/old = 1.5/1 = **1.5**.
From the graph, centre is (–1,–2).
So it is an enlargement, centre (–1,–2), scale factor 1.5.

Now learn how to use your knowledge

Transformations

10 minutes

Use your knowledge

1 Describe the transformations which map

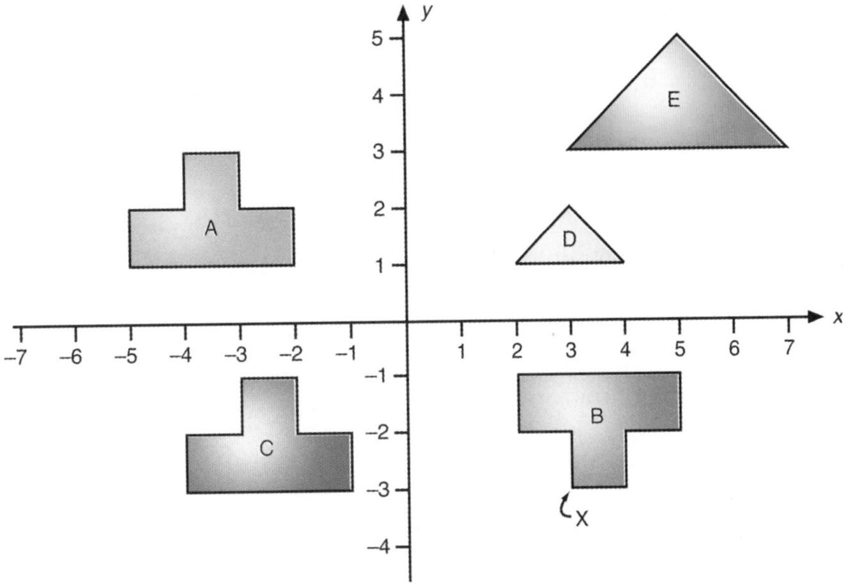

a A to B. Hints 1/2

b C to A. Hints 3/4

c D to E. Hints 5/6

d Describe two successive transformations which move Shape A to Shape B. Hints 7/8

.... then ....

e Write down the coordinates of the point X on Shape B, after Shape B has undergone a translation $\begin{pmatrix}-4\\2\end{pmatrix}$. Hints 9/10/11

✔ Hints and answers follow

Transformations

Hints

1 Use tracing paper.

2 Remember which things you need to describe this kind of transformation.

3 The shape just moves – what is this called?

4 How do we write what has happened to the shape?

5 What is it called when a shape changes size?

6 Remember how to find the two things you need for this transformation.

7 We are looking for one transformation after another.

8 Think of mirrors.

9 What does translation mean?

10 Draw Shape B after the translation.

11 Write down the coordinates of where X is now.

Answers

1 a) Rotation of 180° about (0,0). b) Translation $\binom{-1}{4}$. c) Enlargement scale factor 2 about (1,−1). d) Either 'reflection in x-axis' then 'reflection in y-axis' or other way round. e) Coordinates (−1,−1).

Sequences

10 minutes

Test your knowledge

1 Write down the next two terms of each of the following sequences:

- ***a*** 6, 10, 14, 18, …
- ***b*** 30, 23, 16, 9, …
- ***c*** 3, 6, 12, 24, …
- ***d*** 1, 4, 9, 16, …

2 Find the thirtieth term of the sequence whose n^{th} term is $n^2 + 2n - 1$.

3 Consider the sequence 2, 5, 8, 11, …
Which of the following is the correct formula for its nth term?

- ***a*** $n^2 + 1$
- ***b*** $n + 1$
- ***c*** $3n - 1$.

4 ***a*** What is the formula for the nth term of the sequence 1, 3, 5, 7, … ?

b What is the formula for the nth term of the sequence 0, 3, 8, 15, … ?

Answers

1 a) 22, 26. b) 2, –5. c) 48, 96. d) 25, 36.
2 959. **3** c. **4** a) $2n - 1$. b) $n^2 - 1$.

✓ If you got them all right, skip to page 35

Sequences

30 minutes

Improve your knowledge

1 A sequence

A sequence is a row of numbers that follow some pattern.

5, 8, 11, 14, 17, ... The PATTERN is ADD 3.

1, 2, 4, 8, 16, ... The PATTERN is TIMES BY 2.

1, 4 ,9, 16, 25, ... The PATTERN is SQUARE NUMBERS.

are all sequences.

Questions will ask you to do one of three things: work out the next couple of terms, use a formula you are given or work out a formula for the nth term.

Working out the next couple of terms

The easiest way to do this is to spot the pattern!
What are the next two terms for each of the sequences given above?
The answers are:

- First sequence: 20, 23.
- Second sequence: 32, 64.
- Third sequence: 36, 49 (square numbers are 1×1, 2×2, 3×3, ...).

2 Using and checking the formula

The idea of a formula is that it will allow you to work out – say – the hundredth term without having to work out all of the previous 99.

The formula will have 'n' in it – you substitute the number of the term you want for n. So if you wanted the fiftieth term, you'd put in 50 instead of n.

Example 1 The nth term of a sequence is given by $2n^2 - 3$.
What is the fourteenth term?

We must put in n = 14. So the fourteenth term is $2 \times 14^2 - 3 = 389$.

You may also have to check whether a formula works. This is the same idea! You put in a value for n, and see if the answer you get agrees with what you expect.

Example 2 Tarun is given the sequence 2, 4, 8, 16, … He suggests that the formula for its nth term is 2n. Is he right?

We must check whether the formula works by checking each term.

From the formula:

the first term should be $2 \times 1 = 2$ ✓
the second term should be $2 \times 2 = 4$ ✓
the third term should be $2 \times 3 = 6$ ✗

So the formula doesn't work.

3 Finding the nth term

This is about finding the formulas we were working with before.

You either have to guess them or work them out in a systematic way. The method we are going to use here works when you have to **add** or **subtract** the same number to get from one term to the next.

So it would work for

8, 5, 2, –1, … and 21, 26, 31, 36, …

But not for

0, 3, 8, 15, 24, … or 1, 3, 9, 27, …

This is the method.

The formula will look like (number)n + number, which we will write as:

An + B

where A and B are the numbers we have to find.

The number in front of the 'n' – which we have called A – will be the number we are adding or subtracting. B is the number which comes before the first term.

Example	Find the nth term of the following sequences: a) 8, 10, 12, 14, ...
Step 1	We are adding 2, so we put A = 2.
Step 2	Number before the first term is 6, so put B = 6.
Step 3	Write down formula: n^{th} term = 2n + 6.
	b) 20, 17, 14, 11, ...
Step 1	We are subtracting 3, so we put A = –3.
Step 2	Number before the first term is 23 so put B = 23.
Step 3	Write down formula: n^{th} term = –3n + 23.

Let's check the formula in part b). According to the formula, the fourth term = $-3 \times 4 + 23 = 11$. That's the fourth term in the sequence. So we are correct. ✓

4 What happens if you can't use this method?

We have to guess!

This is usually easy – for example, if your sequence is the square numbers (1, 4, 9, 16, ...), then it is $\mathbf{n^2}$.

Your best bet is to look out for similarities to squared (or cubed) numbers, for example if your sequence is 2 more than cubed numbers, it will be $\mathbf{n^3 + 2}$ (cubed numbers are $1 \times 1 \times 1 = 1$; $2 \times 2 \times 2 = 8$, etc.)

Example	Find the nth term of the sequence: 2, 8, 18, 32, ...

Look for a similarity to square numbers.

Square numbers are 1, 4, 9, 16, ...

Our sequence is **double** this.

So our formula is $\mathbf{2n^2}$.

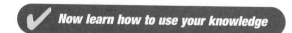

✓ *Now learn how to use your knowledge*

Sequences

10 minutes

Use your knowledge

1 Rahul and Ashley are making patterns with matches, as shown.

1st pattern

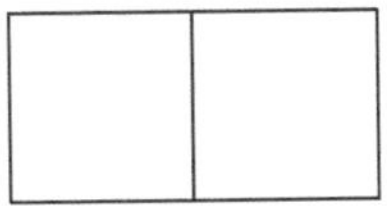

2nd pattern

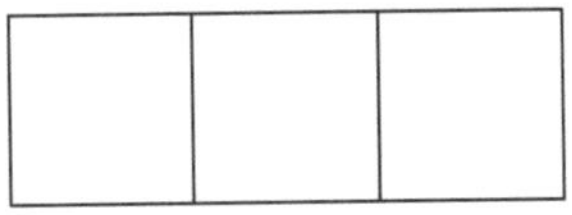

3rd pattern

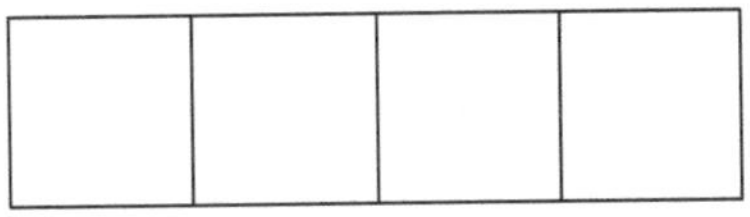

4th pattern

a Draw the fifth pattern. **Hint 1**

b Work out how many matches there are in the sixth pattern, but **do not draw it**. **Hints 2/3/4**

Ashley suggests that the number of matches in the nth pattern is 4n. He explains why he thought this: 'There are n squares in the nth pattern, and you need 4 matches for each square. So the total number of matches must be $4 \times n = 4n$'.

c Explain why Ashley is wrong without doing any calculations. **Hint 5**

Rahul says: 'You need to add 3 matches each time you add a square, so it must be 3n'.

d Show that Rahul is wrong too, by checking his formula. **Hints 6/7**

e Work out what the formula for the nth term should really be. **Hints 8/9/10/11/12**

✔ **Hints and answers follow**

Sequences

Hints

1 How many squares will there be in the next pattern?

2 Try to make a sequence out of the numbers of matches needed.

3 Count up how many matches there were in each of the other patterns and write it out as a sequence.

4 What is the pattern in your sequence?

5 Do you need 4 **new** matches for each square?

6 Check out the formula.

7 Does it work for the first term?

8 Write down the sequence of numbers.

9 Can we use our method?

10 What are we adding or subtracting?

11 What formula do we use?

12 Remember to substitute in.

Answers

1 a) 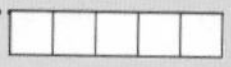 . b) The numbers so far are: 4, 7, 10, 13, 16. So for the next pattern we'll need 19. c) Each time you add a new square, you only add three new matches (or – alternatively – matches are shared between squares). d) If we try Rahul's formula for the first term, we get $3 \times 1 = 3$. But there are really 4! e) We can use the method on this. First look at what we are adding – we are adding 3. So $A = 3$. Now find the term before the first one – it's 1, so $B = 1$. So the formula is $3n + 1$.

Graphs

10 minutes

Test your knowledge

1 Complete the following table and hence draw the graph of $y = 8 - x^2$.

x	–3	–2	–1	0	1	2	3
x^2							
$8 - x^2$							

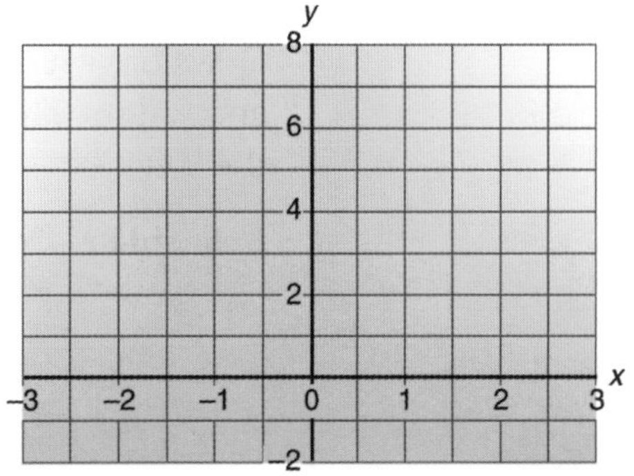

2 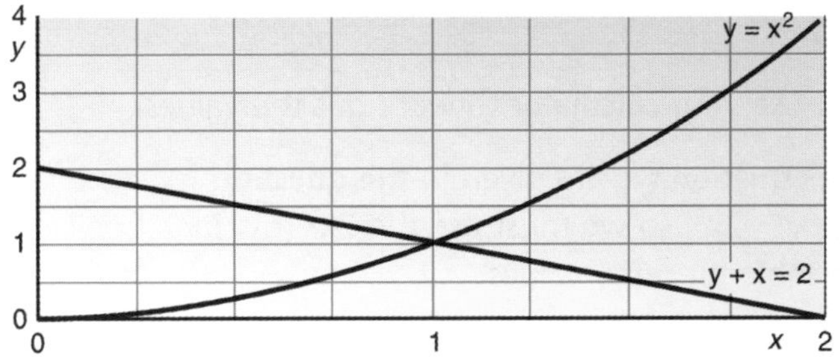

The diagram shows the graphs of $y = x^2$ and $y + x = 2$.

Use the graph to find the solution of the simultaneous equations $y = x^2$ and $y + x = 2$.

3 Find the equation of this line.

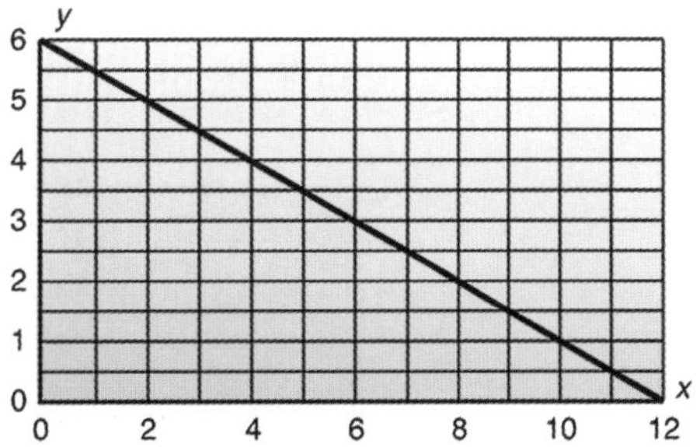

Answers

1 y values are: –1, 4, 7, 8, 7, 4, –1. Graph should be a smooth curve.
2 $x = 1, y = 1$. **3** $y = -0.5x + 6$.

✔ If you got them all right, skip to page 41

Graphs

30 minutes

Improve your knowledge

1 Drawing graphs

Questions on drawing graphs will give you an **equation** connecting y and x, and will usually give you a table with some x values in, and space to work out the y values.

Example Complete the following table and hence draw the graph of $y = 2x^2$.

x	–3	–2	–1	0	1	2	3
x^2	9	4	1	0	1	4	9
$2x^2$	18	8	2	0	2	8	18

First use your calculator to work out the values!

The equation we are given in the question says $y = 2x^2$, so we must use the $2x^2$ row in the table for our y values.

Now draw the graph! First you must work out a scale if the question hasn't done it for you.

Your scale MUST go up in equal jumps. Use easy numbers – like 2 to a square. Try and use most of the graph paper.

Mark each point clearly using a cross. The middle of the cross must be exactly on the point.

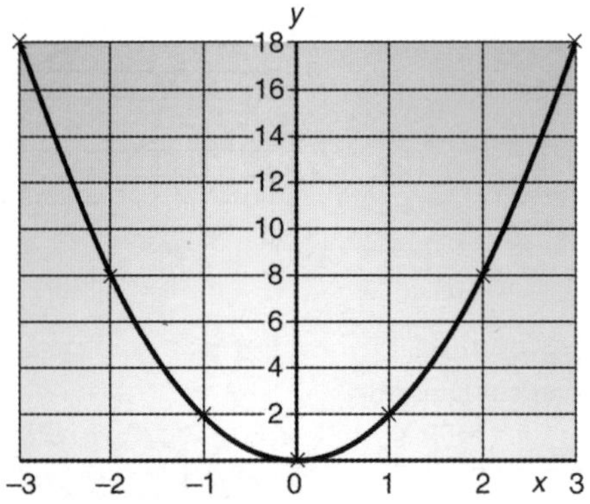

Look at what the points look like! If it is a straight line, use a ruler to join the points up. If it is a curve, join the points up smoothly – no wiggly bits or gaps!

It is possible that the question may not give you a table. If it doesn't, don't worry – make your own!

It's easier to draw a curve if your hand is INSIDE the curve, not outside.

2 Using graphs to solve equations

The key idea here is that: **the solution of two equations is at the point where their graphs cross.**

The question will guide you through drawing the graphs – you just have to realise the importance of that crossing point.

Example The diagram shows the graph of the equation 3x + 4y = 12.

a Fill in the table below for the equation y = x – 0.5.

x	0.5	1.0	1.5	2.0	2.5	3.0	3.5	4.0
x – 0.5	0.0	0.5	1.0	1.5	2.0	2.5	3.0	3.5

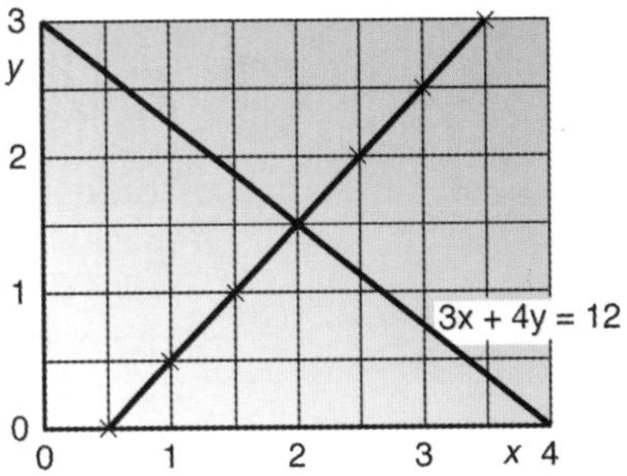

b Draw the line y = x – 0.5 on the same graph.

c Hence write down the solution of the simultaneous equations

3x + 4y = 12
y = x – 0.5.

This is the point where the lines cross.
You need its x value and its y value – so read off **both** axes!

We get x = 2 and y = 1.5.

3 Finding the gradient of a line

The gradient of a line tells you how steep it is – a big gradient means a steep line!

Lines that slope this way have a positive gradient.

Lines that slope this way have a negative gradient.

You find the gradient using the formula

$$\text{Gradient} = \frac{\text{Change in y}}{\text{Change in x}}$$

You find these by drawing a triangle on the graph – the vertical side gives you the change in y and the horizontal side gives you the change in x. Remember to draw the biggest triangle you can.

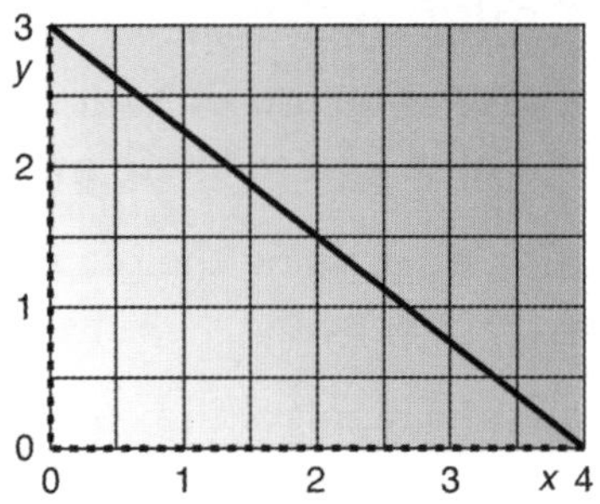

If we want to find the gradient of this graph we could use the triangle shown.

Change in y is 3.

Change in x is 4.

So gradient = $\frac{-3}{4}$

Gradient is minus because of the way the line slopes.

4 Finding the equation of a line

You can find the equation of the line if you are given its graph!

The equation looks like y = (number)x + (number).
This is often written as y = mx + c where m and c stand for the numbers that we have to find out.

You need to remember that:

- m is the gradient of the line
- c is where the line crosses the y axis (called the intercept).

So to find the equation, you need to:

- find the gradient
- see where it crosses the y axis
- put them both into y = mx + c.

Example To find the equation of this line, first find the gradient.

Using the triangle, we get gradient = $\frac{7.5}{10}$ = 0.75

So we put m = 0.75.

Check it's sloping in the positive direction.

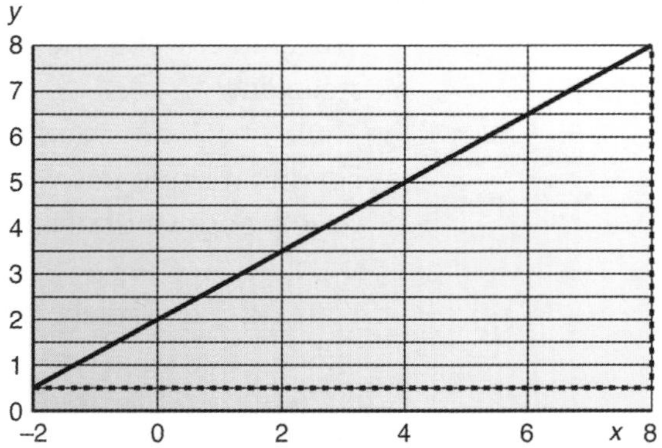

Now look for where the line crosses the y axis. It crosses at 2.0, so c = 2.0.

Now put the numbers into the formula y = 0.75x + 2.0.

Now learn how to use your knowledge

Graphs

10 minutes

Use your knowledge

1 Rearrange each of the following equations to make y the subject. Hints 1/2

a 2y = x + 3.

b x + y = 9. Hints 3/4

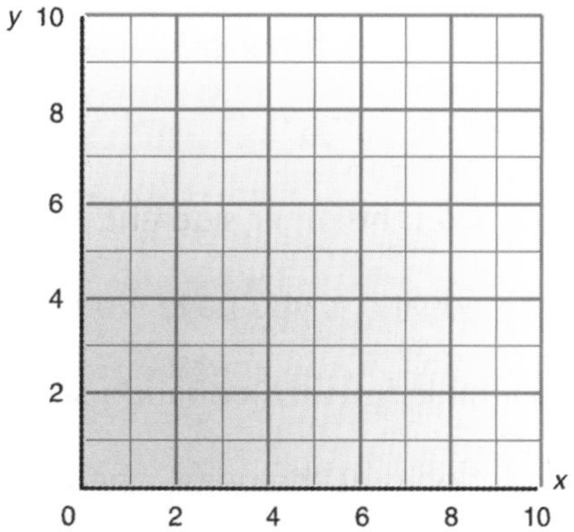

c Plot appropriate graphs to solve the simultaneous equations 2y = x + 3 and x + y = 9 . Hints 5/6

d From your graphs, find the solution of these simultaneous equations. Hints 7/8

2 The temperature in degrees Fahrenheit (F) can be found from the temperature in degrees Centigrade (C) using the following rule.

1 start with C, ***2*** multiply by 9, ***3*** divide by 5, ***4*** add 32.

a Use this information to write down a formula for F in terms of C. Hints 9/10

b Draw a graph with C on the x axis and F on the y axis that you could use to convert Centigrade temperatures between 0 and 35 into Fahrenheit temperatures. Hints 11/12/13

c Use your graph to find the temperature in degrees Centigrade when it is 59 degrees Fahrenheit. Hints 14/15

✓ Hints and answers follow

Graphs

Hints

1 How do we get rid of a number in front?

2 Take the 2 to the other side and divide.

3 Remember it's a '+x'.

4 Put the x on the other side and subtract.

5 Use the equations you have worked out in parts a) and b) with y as the subject.

6 Draw a table for each equation separately using x-values between 0 and 9.

7 The solution is at the point where the lines cross.

8 Remember you need the x value and the y value.

9 Write down what you get as you go along.

10 Write down '9C' first, then write down what you get when you divide this by 5.

11 You need to set up a table with x values going between 0 and 35.

12 You don't need to take x values every 1 – every 5 would be enough!

13 Try using a scale of 1 large square to 5 units on the x-axis and 1 large square to 10 units on the y-axis.

14 Which axis are you starting from?

15 Read across from 59 on the y-axis.

Answers

1 a) $y = x/2 + 3/2$. b) $y = 9 - x$. c) Your graphs will be correct if you have the right answer to part b). d) $x = 5$, $y = 4$. **2** a) $F = \frac{9C}{5} + 32$.

b) Table is:

C	0	5	10	15	20	25	30	35
$\frac{9C}{5} + 32$	32	41	50	59	68	77	86	95

Points join up to give a straight line. c) 15.

Pythagoras and trigonometry

10 minutes

Test your knowledge

1 ***a*** Find the length of the other side of this triangle.

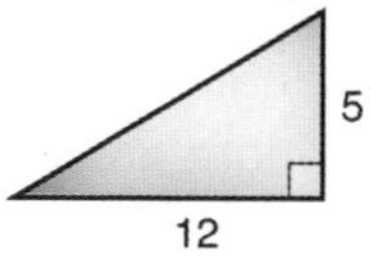

b Find the length of the other side of this triangle, giving your answer correct to 3 decimal places.

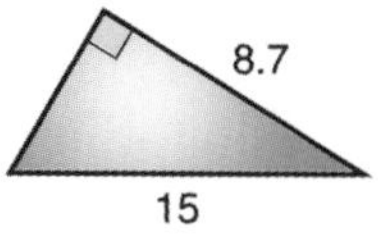

2 ***a*** Find the length of the side marked 'X' in this triangle, giving your answer correct to 3 decimal places.

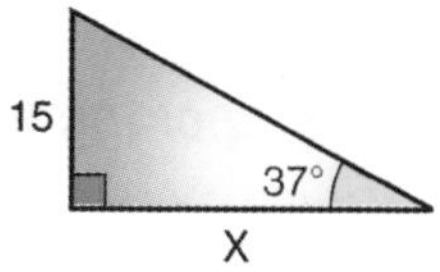

b Find the length of the side marked 'X' in this triangle, giving your answer correct to 3 decimal places.

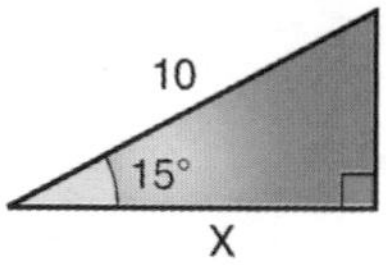

3 Find the size of the angle marked 'x' in this triangle, giving your answer in degrees to 1 decimal place.

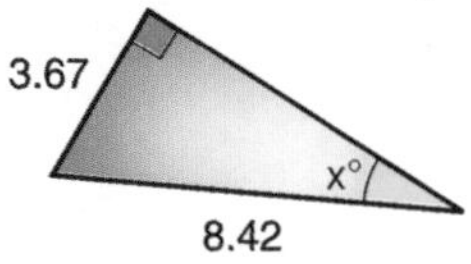

Answers

1 a) 13. b) 12.219. **2** a) 19.906. b) 9.659.
3 25.8°.

✔ If you got them all right, skip to page 48

Pythagoras and trigonometry

30 minutes

Improve your knowledge

1 Pythagoras

Pythagoras said that for a right-angled triangle $\mathbf{c}^2 = \mathbf{a}^2 + \mathbf{b}^2$ where c is the longest side of the triangle (opposite the right angle) and a and b are the other sides.

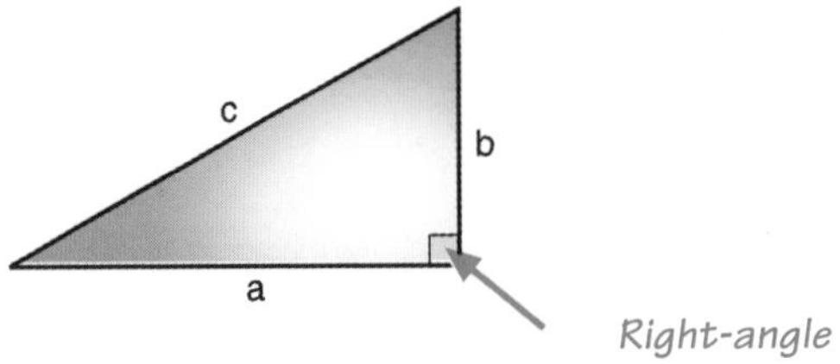

There are two sorts of Pythagoras questions.

1 Looking for the longest side

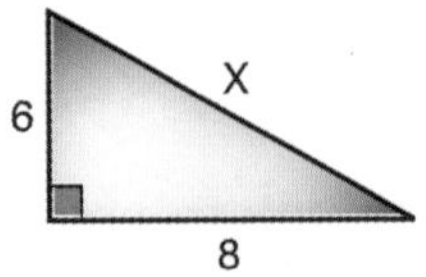

$x = \sqrt{8^2 + 6^2}$

Bigger number first — *Smaller number next*

Looking for the longest side means ADD.

$x = \sqrt{100}$

$x = 10.$

2 Looking for one of the other sides

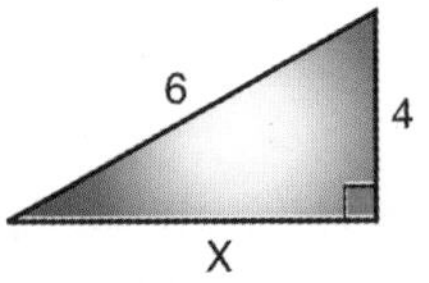

$x = \sqrt{6^2 - 4^2}$

Bigger number first — *Smaller number next*

Looking for one of the other sides means TAKE AWAY.

$x = \sqrt{20}$

$x = 4.47\underline{2}1...$

$x = 4.47$ (2 d.p.).

Remember that we use Pythagoras when we have two sides and want to find the other one.

Trigonometry

Trigonometry is about the angles and sides of triangles. Before we can do any trigonometry, we have to be able to label the sides of a triangle.

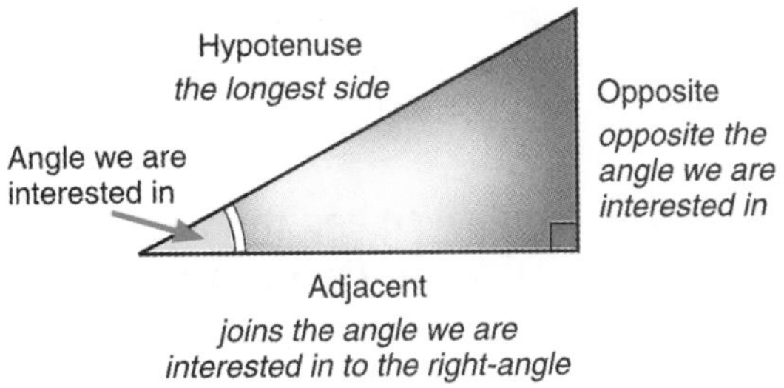

The HYPOTENUSE is always the longest side. But we don't know which side is the opposite and which is the adjacent until we decide on the angle we are looking at.

It doesn't matter which way up the triangle is!

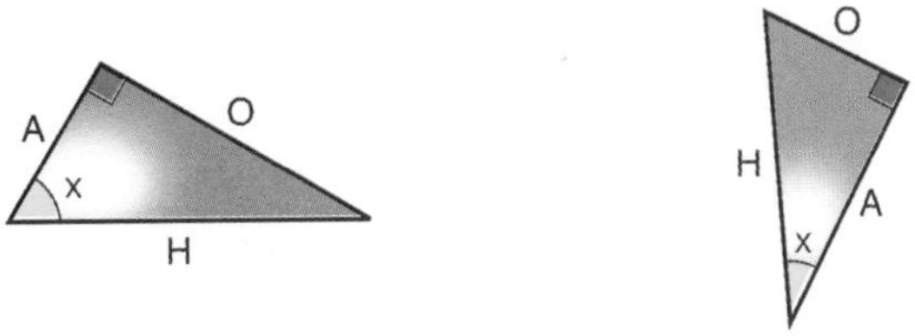

There are three special words we use in trigonometry – **sine** (shortened to **sin**), **cosine** (**cos**) and **tangent** (**tan**).

You will see them on your calculator and on the Formula Sheet in the exam. This is what they mean:

$$\text{Sin} = \frac{\text{Opposite}}{\text{Hypotenuse}} \qquad \text{Cos} = \frac{\text{Adjacent}}{\text{Hypotenuse}} \qquad \text{Tan} = \frac{\text{Opposite}}{\text{Adjacent}}$$

To help remember it, we use the magic word **SOHCAHTOA**.

Can you work out what it means? The answers are given below.

Trigonometry questions will always want you to find the length of a side or the size of an angle. There are three types of question – two kinds where you are finding the length of a side and one where you are finding the size of an angle.

Answers

SOH means that sine is opposite over hypotenuse,
CAH means that cos is adjacent over hypotenuse and
TOA means that tan is opposite over adjacent.

2 Finding a side

Type 1 – letter on top

Example Find the side X in the triangle shown.

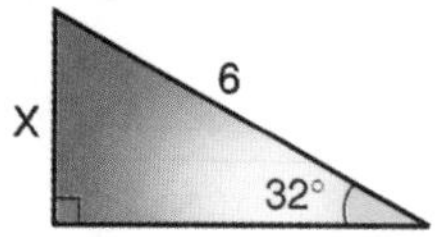

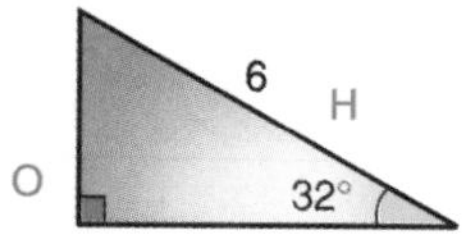

1 Label the side you have been given and the side you want to find out with Opposite, Hypotenuse or Adjacent. **Don't label the other side! You are not interested in it!**

2 Look at SOHCAHTOA. Where are the two letters you've got next to each other?

We've got O and H. So we need to use SOH – that's sin.

3 Say what you see! Write the formula down and put the numbers you know into it.

$$\sin = \frac{\text{Opposite}}{\text{Hypotenuse}}$$

$$\text{So } \sin 32° = \frac{X}{6}$$

4 Find X – if the letter is on the top, we **times**.

$X = 6 \times \sin 32°$

So $X = 3.18$ (2 d.p.).

Remember – letter on the top means TIMES.

Type 2 – letter on bottom

Example Find the side Y in the triangle shown.

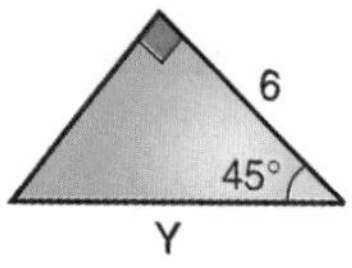

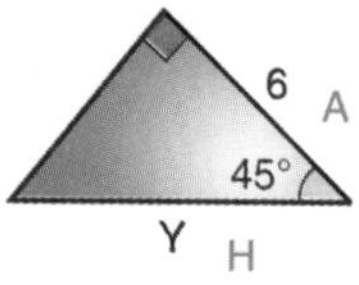

Steps 1, 2 and 3 are exactly the same!

1 Label.

2 Find where you are in SOHCAHTOA.
A and H tells you it is CAH – so use cos.

3 Write down your formula and put numbers into it.

$$\cos = \frac{\text{Adjacent}}{\text{Hypotenuse}}$$

$$\text{So } \cos 45^\circ = \frac{6}{Y}$$

4 We need to find Y now.
The letter Y is on the bottom and the algebra looks a bit nasty!
A quick and easy way is just to **swap** what's on the bottom with the top on the other side – the boxes swap places!

$$\boxed{\cos 45^\circ} = \frac{6}{\boxed{Y}}$$

$$Y = \frac{6}{\cos 45^\circ}$$

So Y = 8.49.

Remember
– letter on bottom
means SWAP.

3 Finding an angle

Example Find the angle X in the triangle shown.

Steps 1, 2 and 3 are just the same!

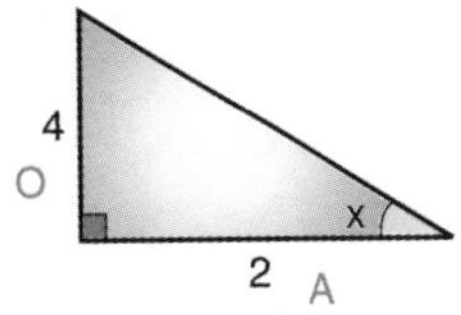

1 Label.

2 Find where you are in SOHCAHTOA.

O and A means we need TOA – so tan.

3 Write down the formula.

$$\tan = \frac{\text{Opposite}}{\text{Adjacent}}$$

$$\tan x = \frac{4}{2}$$

4 To find the angle we have to use **inverse** sine, cosine or tangent (written $\sin^{-1}$, $\cos^{-1}$ and $\tan^{-1}$).
You get these by pressing the SHIFT or 2ND key on your calculator before the sin, cos or tan key.

tan x = 2
we use inverse tan
$x = \tan^{-1} 2$
$x = 63.4^\circ$.

Remember
– finding an angle
means INVERSE.

✓ Now learn how to use your knowledge

Pythagoras and trigonometry

10 minutes

Use your knowledge

1 Anita and Sam are standing together at corner X of the school playing field.

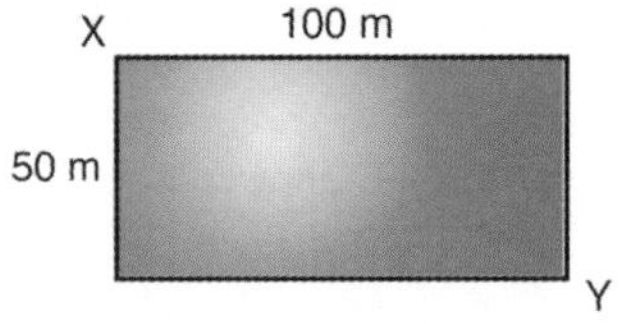

Sam walks around the edge of the field to get to corner Y, but Anita walks directly across the field. How much further does Sam walk than Anita?

Hints 1/2/3

2 The diagram shows a flagpole BCD held up by two ropes, AB and CE.

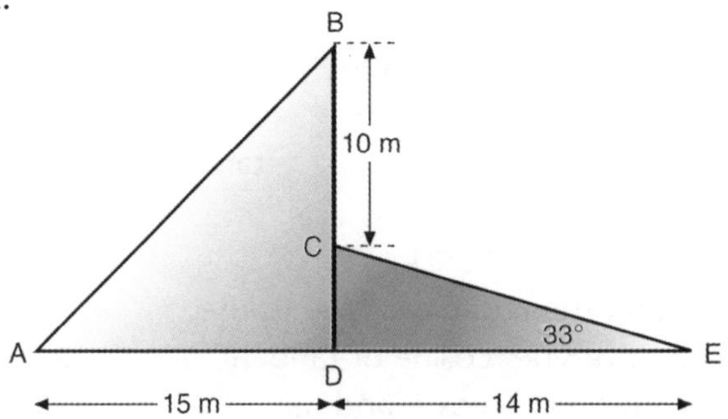

a Find in metres the distance of C above the ground. Hints 4/5

b Find the length in metres of rope AB. Hints 6/7/8

c Find the angle BÂD. Hints 9/10/11

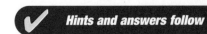

Hints and answers follow

Pythagoras and trigonometry

Hints

1. Work out how far each of them walk.
2. Find a triangle with the distance Anita walks as one of its sides.
3. Use Pythagoras on a triangle made up of two sides of the field and its diagonal.
4. Look at triangle CDE.
5. Use tan to work out CD.
6. Look at triangle ABD.
7. What is the total length of BD?
8. Use Pythagoras to find length AB.
9. Angle BÂD means the angle A.
10. Use triangle ABD.
11. Use tan to find the angle.

Answers

1 Anita walks $\sqrt{(50^2 + 100^2)}$ = 111.8 m. Sam walks 100 + 50 = 150 m. So Sam walks 38.2 m more. **2** a) CD/14 = tan33° so CD = 9.09 m. b) We use the triangle ABD. We now know length BD = 10 + 9.09 = 19.09 m. Using Pythagoras, AB = $\sqrt{(19.09^2 + 15^2)}$ = 24.28 cm. c) Using triangle ABD again, tanA = 19.09/15 so A = 51.8°.

Shapes

10 minutes

Test your knowledge

1 Calculate the area of the following shapes.

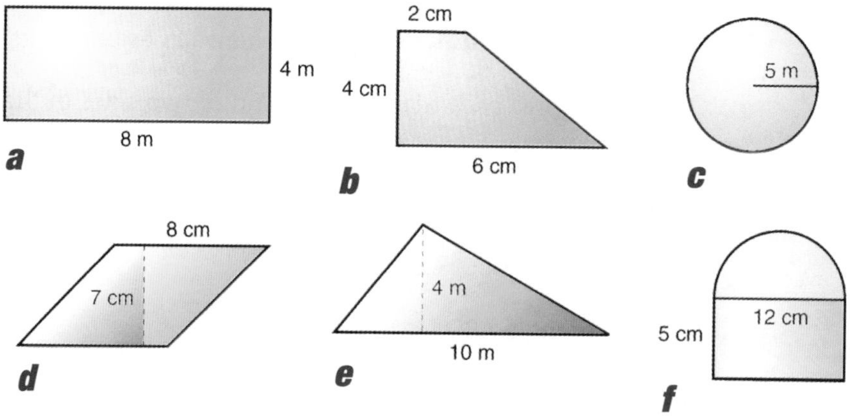

2

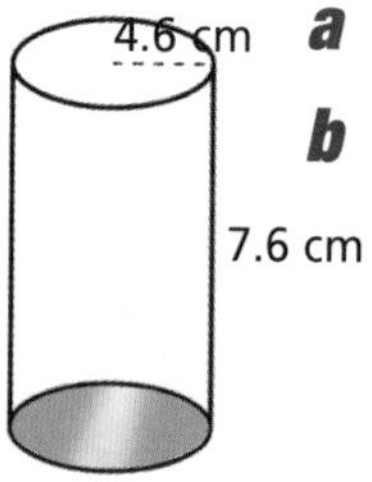

a What is the circumference of the shaded circle?

b Calculate the volume of the shape, giving your answer to 3 s.f.

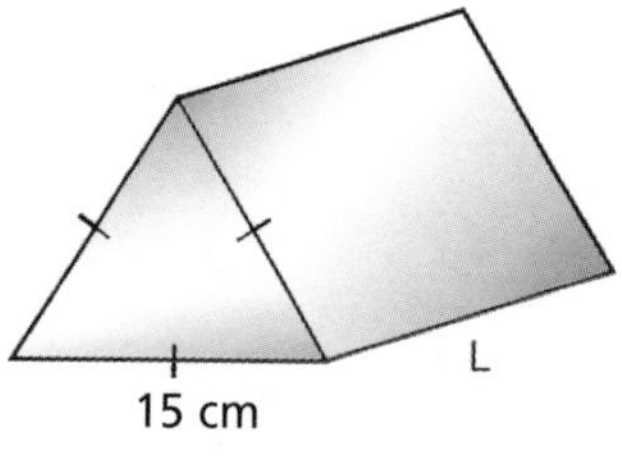

c Calculate the area of the triangle.

d The volume of this triangular prism is 1000 cm^3.

Calculate its length, L.

Answers

1 a) 32 m^2. b) 16 cm^2. c) 78.54 m^2. d) 56 cm^2. e) 20 m^2. f) 116.549 cm^2. **2** a) 28.9 cm. b) 505 cm^3. c) 97.43 cm^2. d) 10.26 cm.

✔ If you got them all right, skip to page 54

Shapes

30 minutes

Improve your knowledge

Shapes – names, areas

You need to recognise the following shapes and be able to calculate their areas.

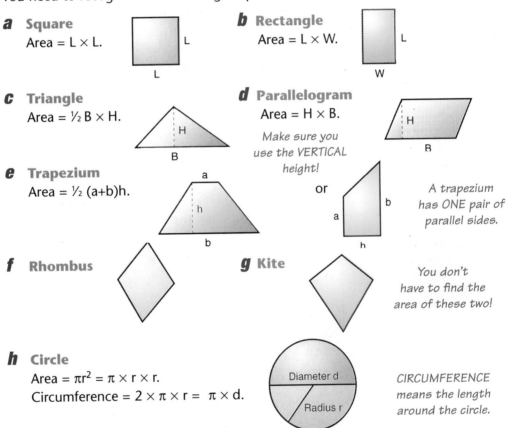

a **Square**
Area = L × L.

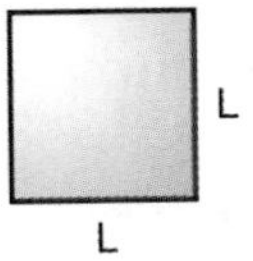

b **Rectangle**
Area = L × W.

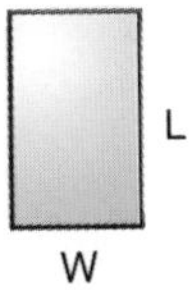

c **Triangle**
Area = ½ B × H.

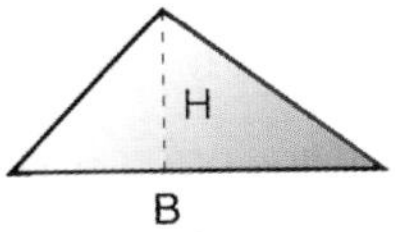

d **Parallelogram**
Area = H × B.

Make sure you use the VERTICAL height!

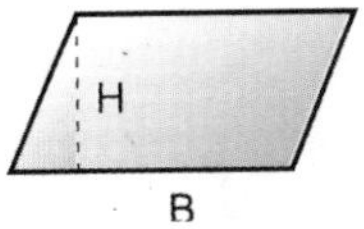

e **Trapezium**
Area = ½ (a+b)h.

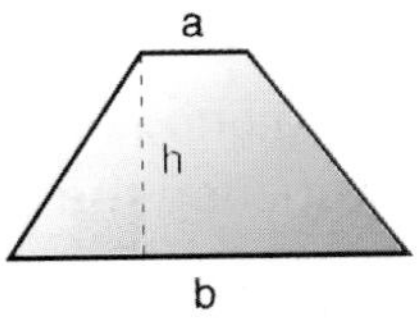

or

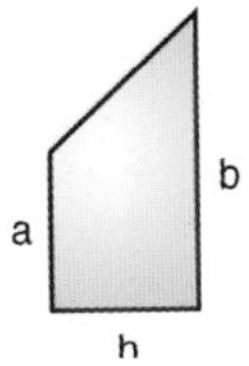

A trapezium has ONE pair of parallel sides.

f **Rhombus**

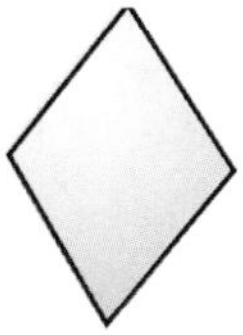

g **Kite**

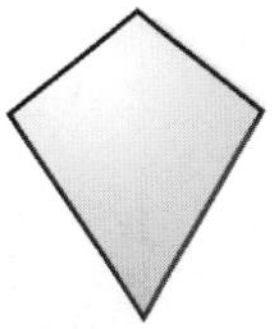

You don't have to find the area of these two!

h **Circle**
Area = $\pi r^2 = \pi \times r \times r$.
Circumference = $2 \times \pi \times r = \pi \times d$.

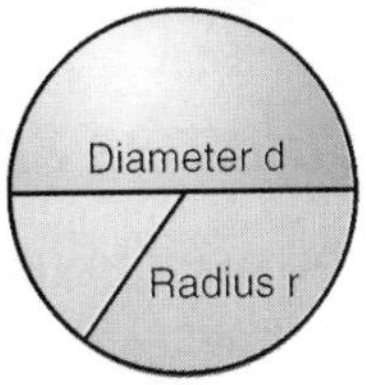

CIRCUMFERENCE means the length around the circle.

i Other shapes: a **Pentagon** has five sides,
a **Hexagon** has six sides and an **Octagon** has eight sides.

1 Using area formulas

Sometimes you may have to use the area formulas for shapes that aren't exactly like the ones on the list.

Example Calculate the area and the perimeter of this shape.

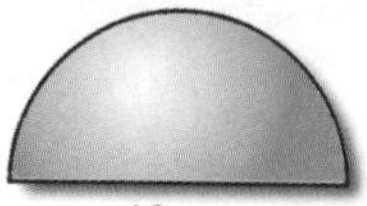

The shape is half a circle. So we must look at the formulas to do with circles.
Area of circle = $\pi \times r \times r$. This is half a circle so area = $\frac{1}{2} \times \pi \times r \times r$.
We know the diameter = 16 cm. So the radius, r = 8 cm.
Check your units cm^2 or m^2 for an area.
So the area = $\frac{1}{2} \times \pi \times 8 \times 8 = 100.5\ cm^2$.

The perimeter is made up of a curved bit and a straight bit.
The straight bit is 16 cm.
PERIMETER is the length round the outside.
The curved bit is half the circumference of a circle = $\frac{1}{2} \times \pi \times d$ = $\frac{1}{2} \times \pi \times 16 = 25.13$ cm.
So the total perimeter = straight bit + curved bit = 16 + 25.13 = 41.13 cm.

Example Find the area of this shape.

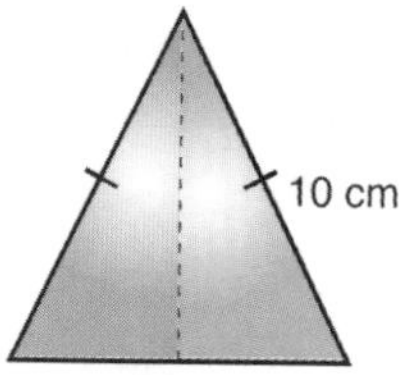

We don't know what the vertical height is!
(Make sure you understand why it isn't 10 cm.)
So we must find the vertical height before we can find the area.
We only know how to find lengths on **right-angled** triangles, so we must chop the triangle in half first.

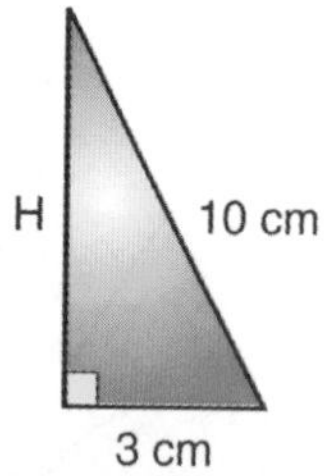

Use Pythagoras $H = \sqrt{10^2 - 3^2}$
$= \sqrt{91} = 9.539$ cm.

SHORTER side so SUBTRACT.

Now we can find the area.
Area = $\frac{1}{2} \times B \times H = \frac{1}{2} \times 6 \times 9.539 = 28.6\ cm^2$.

2 Volumes and 3-D shapes

You need to recognise the following shapes and be able to calculate their volumes.

a Cylinder

Volume = $\pi r^2 h = \pi \times r \times r \times h$.

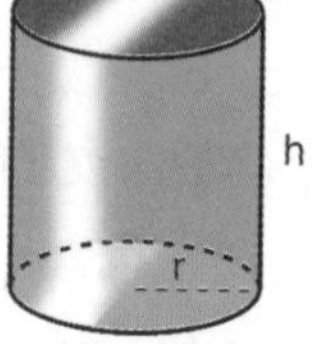

b Prism

Volume = (cross-sectional area) × (length)
= A × L.

A prism can have any shape as its cross-section.

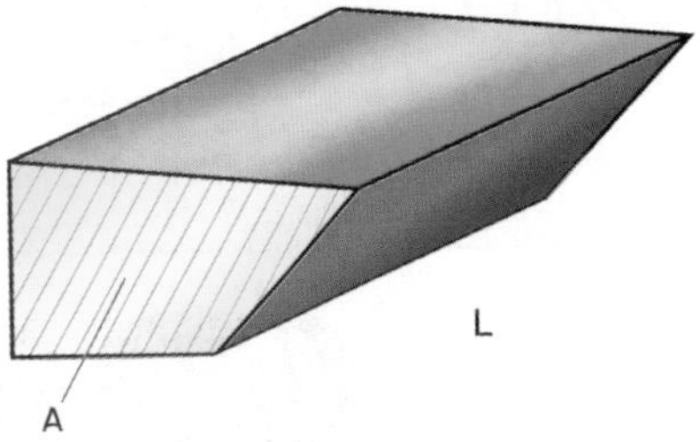

Example Calculate the volume of this soapbox.

Step 1 Write the formula down.
Volume of prism = A × L.

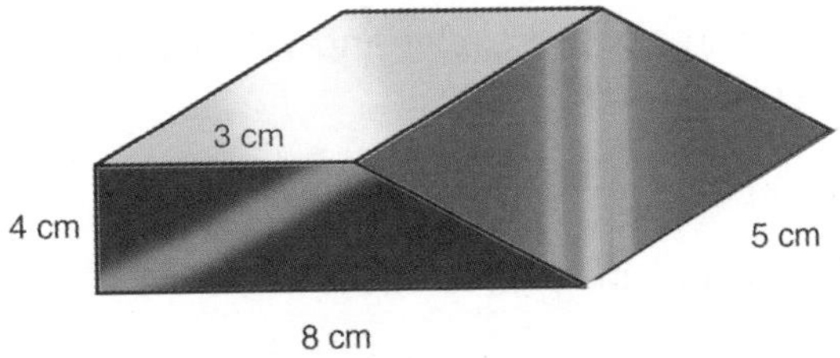

Step 2 Work out A – that's the area of the trapezium at the front.
Area of trapezium = $\frac{1}{2} \times (a + b) \times h$
$= \frac{1}{2} \times (8 + 3) \times 4$
$= \frac{1}{2} \times 11 \times 4 = 22\ \text{cm}^2$.

Step 3 Put it back in the volume formula.
Volume = A × L = 22 × 5 = 110 cm^3.

Example A cylindrical can of soft drink has a radius of 2.5 cm and has a capacity of 330 cm^3. Calculate

a The area of the top of the can

b The height of the can.

Working this out we get:

a The top of the can is a circle.
So area = $\pi r^2 = \pi \times 2.5 \times 2.5 = 19.634\,95\ldots\ \text{cm}^2$.

b To find the height, we can use the formula V = A × L.
V = 330 cm^3, A = 19.634 95 so: 330 = 19.634 95 × L

$\frac{330}{19.634\,95} = L$ so L = 16.8 cm.

✔ ***Now learn how to use your knowledge***

Shapes

10 minutes

Use your knowledge

1 The diagram shows a cross-section through a box containing 4 cylinders each of radius 8 cm and height 15 cm. On the diagram, W, X, Y and Z are the centres of the circles and WXYZ is a square. Calculate:

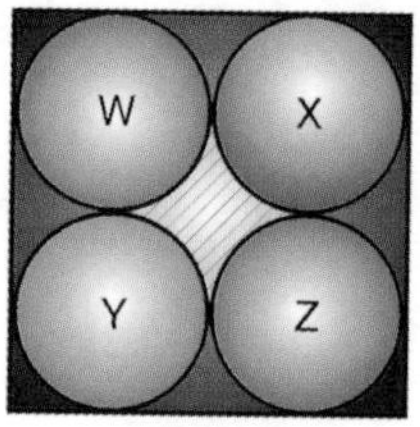

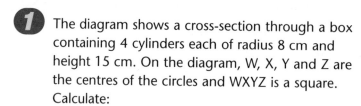

a The circumference of the circle W. *Hint 1*

b The perimeter of the shaded region. *Hints 2/3*

c The area of the shaded region. *Hints 4/5/6*

d The volume of the box. *Hints 7/8*

e The volume of the four cylinders. *Hints 9/10*

2 A cylinder has a base radius 3.6 cm.

a Calculate the area of this base. *Hint 11*

b The volume of the cylinder is 500 ml. Calculate the height of the cylinder. *Hints 12/13/14*

3 The shape below is the cross-section of a triangular prism 20 cm long. Calculate the volume of the prism. *Hints 15/16*

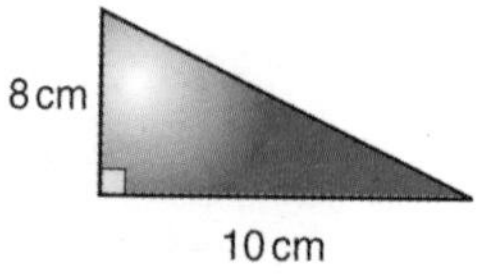

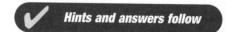

Hints and answers follow

Shapes

Hints

1 Remember the circumference of a circle formula.

2 What does perimeter mean?

3 The perimeter is made up of four curves. What shape do they add up to make?

4 What shape is WXYZ?

5 What do the 4 unshaded regions in WXYZ make?

6 How can you get the shaded area from WXYZ and the unshaded parts?

7 What solid shape will the box be?

8 Look through the question to find the length of the box.

9 What is the formula for the volume of a cylinder?

10 All four cylinders are the same!

11 What is the formula for the area of a circle?

12 500 ml = 500 cm^3.

13 Write down the formula for the volume of a cylinder.

14 Put in the things you know.

15 Work out the cross-sectional area.

16 It's a prism!! What's the formula?

Answers

1 a) $C = 2 \times \pi \times 8 = 50.265...$ cm. b) P = circumference of circle = 50.265... cm.
c) Area of square WXYZ = $16 \times 16 = 256$ cm^2. Area of circle (unshaded region) = $\pi \times 8 \times 8 = 201.0619...$ cm^2. Shaded area = $256 - 201.0619... = 54.938...$ cm^2. d) Volume of cuboid = $L \times W \times H$. $L = 4 \times 8 = 32$. So volume = $32 \times 32 \times 15 = 15360$ cm^3. e) Volume of cylinder = $\pi \times 8 \times 8 \times 15 = 3015.9289...$ So volume of 4 cylinders = $4 \times 3015.9289... = 12063.7...$ cm^3.

2 a) Area = $\pi \times 3.6 \times 3.6 = 40.715...$ cm^3. b) $V = A \times L$. $500 = 40.715 \times L$ so $L = 500/40.715 = 12.28...$ cm.

3 Area of triangle = $\frac{1}{2} \times 10 \times 8 = 40cm^2$. Volume of prism = $A \times L = 40 \times 20 = 800$ cm^3.

Statistics 1

Test your knowledge

10 minutes

1 The frequency table below shows how marks were awarded in a test.

Marks	2	3	4	5	6	7	8	9	10
Frequency	4	8	7	7	5	4	3	3	1

Calculate:

a The mean mark.

b The median mark.

c The range of marks.

d The modal mark.

2 The following frequency table shows the weights of eggs laid by hens at my local farm.

Weight (W)	$20 < W \le 30$	$30 < W \le 40$	$40 < W \le 50$	$50 < W \le 60$	$60 < W \le 70$
Frequency	7	14	23	18	5

a Estimate the mean weight of the eggs.

b Write down the modal group.

3 The following is a cumulative frequency curve for the marks 50 students scored in a test. Calculate the:

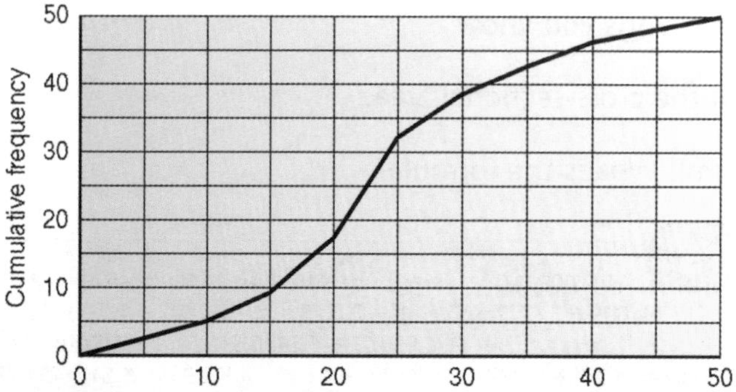

a median.

b upper quartile.

c lower quartile.

d interquartile range.

Answers

1 a) 5.0952... b) 5. c) 8. d) 3.
2 a) $^{3015}/_{67} = 45$. b) $40 < W \le 50$.
3 a) 22. b) 28. c) 18. d) 10.

✔ If you got them all right, skip to page 61

Statistics 1

Improve your knowledge

30 minutes

Averages and ranges

You must learn what these words mean.

Mode is the commonest value.

Median is the **middle** value when you have written the data in order. The formula $\frac{n+1}{2}$ tells you how many places along the list you have to go (n = number of values).

Mean is what you get by adding all the values up and dividing by how many there are.

Range is the biggest value take away the smallest value.

1 Ungrouped frequency tables

This table shows the price of a Mars Bar in 25 different shops:

Price	24	25	26	27	28	29	30
Frequency	1	2	6	7	4	3	2

This is a frequency table.

These add up to 25 = number of shops.

a How many shops charged 29p or more for a Mars Bar?

3 shops charged 29p.
2 shops charged 30p. So 3 + 2 = 5 shops charged 29p or more.

b What is the modal price?

The biggest frequency is 7. This is for the price 27p.
So the mode is 27p.

The mode is the value with the biggest frequency.

c Calculate the mean.

Price (x)	24	25	26	27	28	29	30
Frequency (f)	1	2	6	7	4	3	2
fx	24	50	156	189	112	87	60

Step 1 Label the frequency row f and the price row x.

Step 2 Put an extra row on the bottom of the table – call it fx.

Step 3 Multiply each of the x values by its f value.

Step 4 Use the formula

$$\text{mean} = \frac{(\text{add up the fx values})}{(\text{add up the f values})} = \frac{(24+50+156+189+112+87+60)}{(1+2+6+7+4+3+2)} = \frac{678}{25} = 27.12.$$

d Calculate the median price.

Step 1 Find the total number by adding up the f values

Total is 25.

Step 2 Use the formula $\frac{n+1}{2}$

$\frac{25+1}{2} = 13.$

Step 3 Go along the frequency table and keep adding the frequencies until you hit or go over the number you got from your formula

1, then 3, then 9, then 16 (we've gone over).

Step 4 Look up to see which x value you've got to! This is the median

We've gone as far as 27p – so it's the median.

2 Grouped frequency tables

These are quite like the normal frequency tables above.

The heights of 30 students were measured and recorded in the following frequency table.

Height (cm)	Frequency (f)	Midpoint (x)	fx
$120 \leq h < 130$	1	125	125
$130 \leq h < 140$	2	135	270
$140 \leq h < 150$	11	145	1595
$150 \leq h < 155$	10	152.5	1525
$155 \leq h < 160$	5	157.5	787.5
$160 \leq h < 180$	1	170	170

a Calculate the mean height.

Step 1 Find the mid point for each group by averaging the top and bottom values for that group. Put these in the table as an extra column.

Step 2 Label the frequency column f and the midpoint column x. Make another new column called fx.

Step 3 Fill in the fx column by multiplying each x value by the f value that goes with it.

Step 4 Use the formula (*same as last time*):

$$\text{mean} = \frac{\text{(add up the fx values)}}{\text{(add up the f values)}} = \frac{(125 + \ldots + 170)}{(1+2+ \ldots +1)} = \frac{4472.5}{30} = 149.08 \text{ cm.}$$

b State the modal group.

This is the group with the biggest frequency: $140 \le h < 150$.

3 *Cumulative frequency*

Cumulative frequencies are just a running total of all the frequencies.

Example The masses of 60 apples were recorded on a frequency table.

Mass (g)	Frequency	Cumulative frequency	
$80 \le m < 85$	2	2	*(2)*
$85 \le m < 90$	5	7	*(2 + 5)*
$90 \le m < 100$	23	30	*(2 + 5 + 23)*
$100 \le m < 105$	15	45*	*(2 + 5 + 23 + 15)*
$105 \le m < 110$	11	56	*(2 + 5 + 23 + 15 + 11)*
$110 \le m < 120$	4	60	*(2 + 5 + 23 + 15 + 11 + 4)*

a Insert the cumulative frequencies in the table.

b What does the starred number mean in this table?

It means that 45 apples had a weight of 105 grams or less.

Top value from the class. *Always put this in.*

c Draw a cumulative frequency graph for this data.

To do this, we must plot the **top value** of each class on the **x axis** against the **cumulative frequency** on the **y axis**.

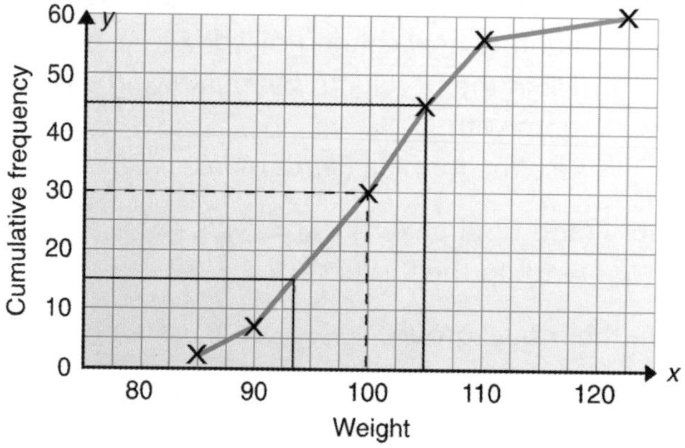

d Using your curve, find the median.

Step 1	Use the median formula for curves, **n/2**: 60/2 = 30.
Step 2	Find this value on the y axis and draw a line across to the curve.
Step 3	Draw a line down to the x axis and read off the value. We get 100. So median is 100.

e Find the interquartile range (IQR).

Finding the quartiles is quite like finding the median.

Step 1	Use the lower quartile formula, $\frac{n}{4}$:	$\frac{60}{4} = 15$
Step 2	Use your graph to find the lower quartile	From the graph we get 93.5
Step 3	Use the upper quartile formula, $\frac{3n}{4}$:	$\frac{3 \times 60}{4} = 45$
Step 4	Use your graph to find the upper quartile	From the graph we get 105
Step 5	Work out the interquartile range by doing Upper quartile – lower quartile	IQR = 105 – 93.5 so IQR = **11.5**.

Remember – n/2 for cumulative frequency, (n + 1)/2 for everything else.

Now learn how to use your knowledge

Statistics 1

Use your knowledge

10 minutes

1 Indiana James was trapped in a cave with many deadly snakes. He decided to get his ruler out and measure them. His results are shown in the frequency table below:

Length (cm)	Frequency	Cumulative frequency
$0 \le l < 40$	14	
$40 \le l < 90$	21	
$90 \le l < 130$	32	
$130 \le l < 180$	26	
$180 \le l < 200$	7	

a Calculate the mean length of the snakes. **Hints 1/2/3**

b Complete the cumulative frequency column in the table. **Hint 4**

c Draw a cumulative frequency graph for the data. **Hints 5/6**

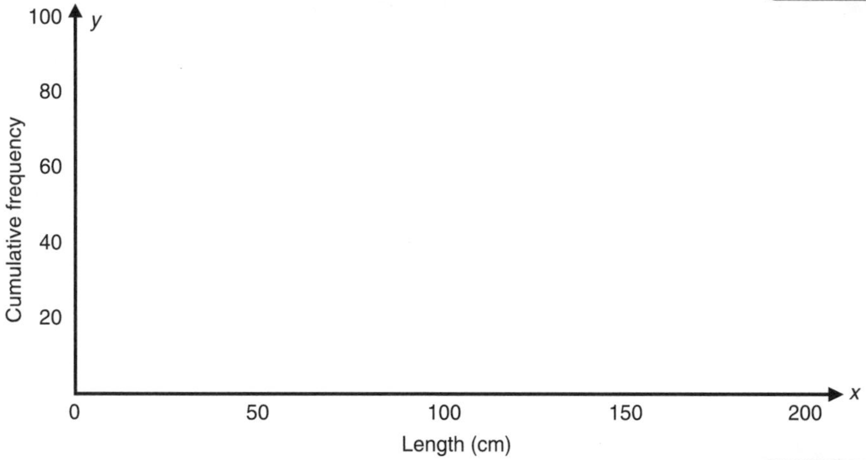

d Use the graph to find the median length. **Hints 7/8/9**

e Snakes longer than 150 cm can strangle you. How many snakes are able to do this? **Hints 10/11/12/13**

Hints and answers follow

Statistics 1

Hints

1 Add a midpoint column.

2 Add an fx column.

3 Mean formula!

4 Keep on adding!

5 Use the table.

6 Remember to plot the endpoints.

7 n/2.

8 What is n?

9 Find n/2 on the y-axis.

10 Since 150 cm is a length, what axis must we look at?

11 Work out how many snakes are smaller than 150 cm first.

12 What does cumulative frequency mean?

13 How can we work out how many snakes are longer than 150 cm if we know how many are shorter?

Answers

1 a) Midpoints: 20; 65; 110; 155; 190. fx values: 280; 1365; 3520; 4030; 1330. Mean = (280 + 1365 + 3520 + 4030 + 1330)/(14 + 21 + 32 + 26 + 7) = 10 525/100 = 105.25 cm. b) 14, 35, 67, 93, 100. c) Curve plotted with C.F.s (y axis) plotted against end points (x axis). d) 109 cm. e) From graph, there are 77 snakes less than 150 cm. So there are 100 – 77 = 23 longer than this.

Statistics 2

10 minutes

Test your knowledge

1 Farmer Giles has a farm with four types of animal.

Type	Number of animals	Angle on Pie Chart
Sheep	180	
Chickens	120	
Cows	99	
Goats	141	

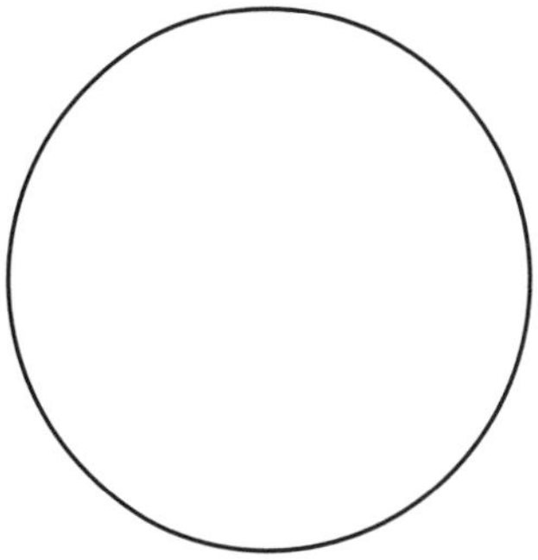

Draw a pie chart to show the different types of animal on the farm.

2 The pie chart (right) shows how Kieran spent his pocket money of £10.80 this week.

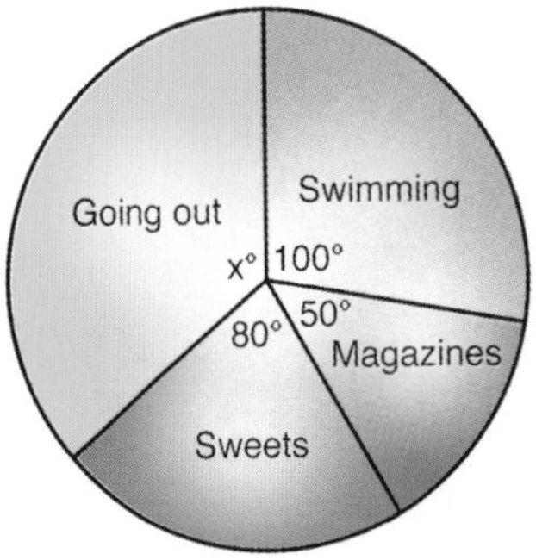

a Calculate the amount of money Kieran spent on sweets.

b Calculate the amount of money Kieran spent on going out.

c Calculate the percentage of the money spent on magazines.

3 The scatter diagram shows the maths marks and physics marks scored by 9 students who took both tests.

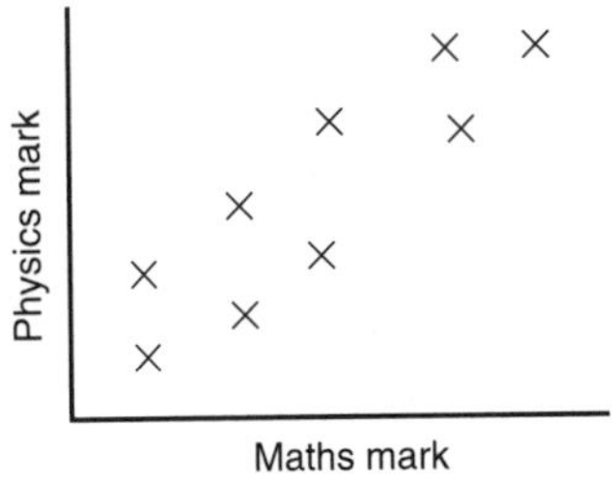

a Draw a line of best fit on the diagram.

b What is the type of correlation between maths marks and physics marks?

Answers

1 Sheep 120°, chickens 80°, cows 66°, goats 94°. **2** a) £2.40. b) £3.90. c) 13.889%. **3** a) Your answer should be a straight line through co-ordinates. See p 66 for example. b) positive.

If you got them all right, skip to page 67

Statistics 2

30 minutes

Improve your knowledge

Pie charts

Pie charts are a way of showing data as a picture.
There are two types of question on pie charts.

The angles in a pie chart must add up to 360°.

1 Drawing a pie chart

270 students were asked which of four subjects they preferred.
The table shows what they said.

Subject	Frequency
Maths	45
Science	30
History	75
Art	120

Draw a pie chart to show this information.

Step 1 Add up all the frequencies. $45 + 30 + 75 + 120 = 270$.

Frequency for maths.

Always 360° for drawing a pie chart.

Step 2 Find the angles.

$$\text{Angle for maths} = \frac{45}{270} \times 360 = 60°.$$

Sum of all the frequencies.

Try working out all the other frequencies!
The answers are given below.

Answers

Science 40°, History 100°, Art 160°.

Step 3 Draw the pie chart.

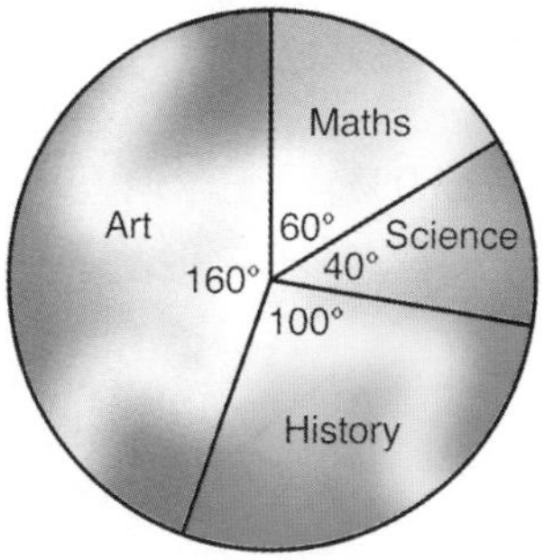

Remember to use your angle measurer and label each part.

2 Doing calculations

300 students were asked what they ate for breakfast.
The pie chart shows their answers.

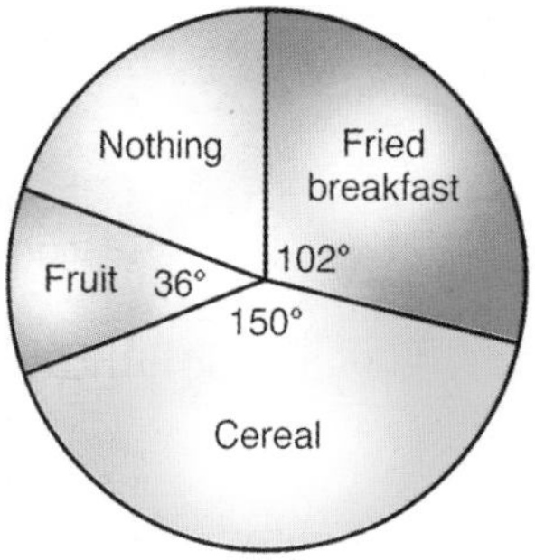

Calculate

a how many students had cereal for breakfast.

b how many students had nothing for breakfast.

c what percentage of students had fruit for breakfast.

Angle for cereal. *Total number of students.*

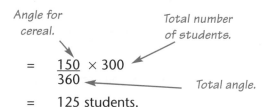

a Number of students $= \dfrac{150}{360} \times 300$ *Total angle.*

$= 125$ students.

We want a number of students. So we times by the total number of students.

b Looks like the same sort of question – BUT ! we are not given an angle for nothing.

So the first thing to do is to find what the angle is.

We must use the fact that the angles add to 360°. So the angle for nothing is $360 - (102 + 36 + 150) = 72°$.

So the number of students who ate nothing $= \dfrac{72}{360} \times 300 = 60$ students.

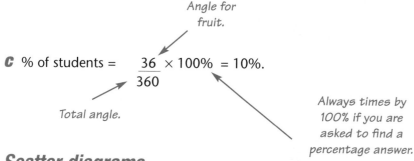

Angle for fruit.

C % of students = $\frac{36}{360} \times 100\% = 10\%$.

Total angle.

Always times by 100% if you are asked to find a percentage answer.

3 Scatter diagrams

Scatter diagrams tell us about how two sets of data (such as the heights and weights of a class of students) are related. They are like graphs, except you just have to plot the points without joining them up.

There are two sorts of relationship. They are called **positive correlation** and **negative correlation**.

Positive gradient so positive correlation.

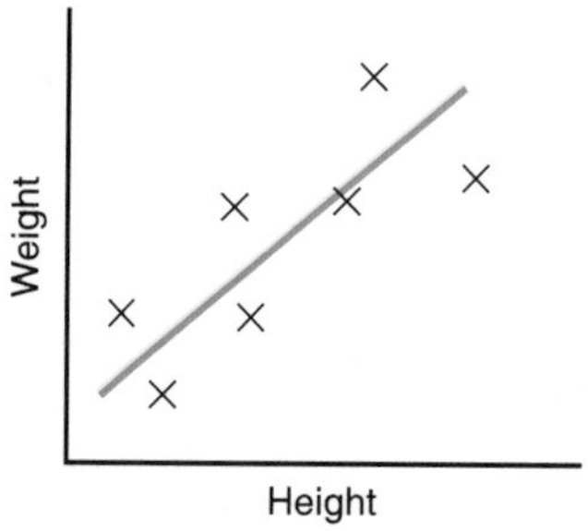

This tells us 'the taller you are the heavier you probably are'.

Negative gradient, so negative correlation.

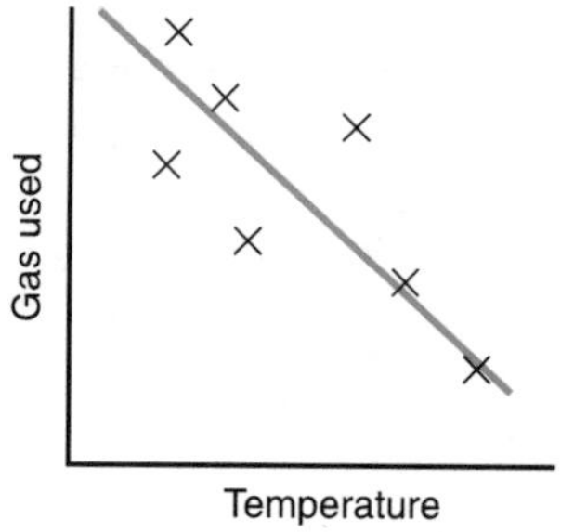

This tells us 'the higher the temperature, the less gas you are likely to use'.

If the points are just scattered around all over the place, there is no correlation.

You can also be asked to draw a **line of best fit**. This is a line that tries to go as close as possible to all the points.

The lines of best fit have been drawn on for the two scatter diagrams above.

✓ *Now learn how to use your knowledge*

Statistics 2

10 minutes

Use your knowledge

1 Amy asked some students how many pets they had at home. She drew a pie chart using her results.

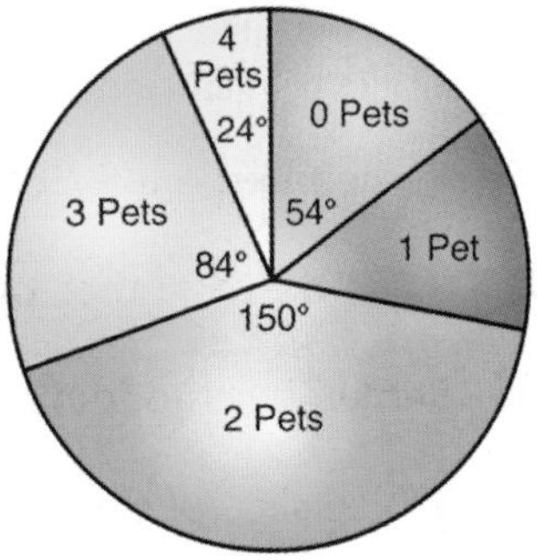

a Calculate the percentage of her friends who have one pet. Hints 1/2/3

b State the modal number of pets. Hints 4/5

c Use the pie chart to work out the missing frequencies in the frequency table. Hints 6/7/8/9

No. of pets	0	1	2	3	4
Frequency	9				

d Hence calculate the mean number of pets. Hints 10/11/12

2

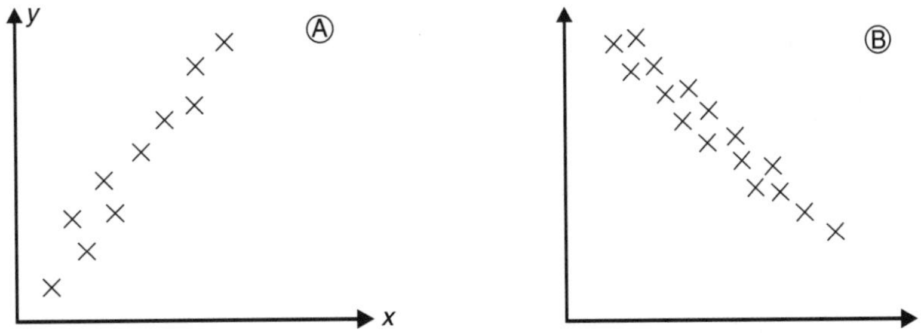

a What type of correlation does scatter diagram B exhibit? Hint 13

b Niall says, 'The more pages in a book the heavier the book will be'. Which scatter diagram could represent this statement? Hint 14

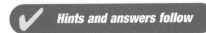

Hints and answers follow

Statistics 2

Hints

1 What is the angle on the pie chart for 1 pet?

2 Remember: angle/total angle.

3 What do you times by to get a percentage answer?

4 Remember mode means the most popular number.

5 Which slice of the pie means the largest area?

6 First work out the total number of students.

7 Remember the formula: Number = (angle for slice)/360° × total number.

8 Use the formula in hint 7 for the people with 0 pets and rearrange to find total number of students.

9 Use the formula for working out numbers to get the rest of the frequencies.

10 Label the rows and add an 'fx' row.

11 How do we get the fx row?

12 What is the formula for the mean for tables?

13 Which way do the scatter points go?

14 What type of correlation is Niall describing?

Answers

1 a) Angle for '1 pet' = 360 – (54 + 24 + 84 + 150) = 48°. So percentage = $\frac{48}{360}$ × 100% = 13.33%. b) 2 pets. c) Number = (angle for slice)/360 × total number. For 0 pets: 9 = 54/360 × total. So 9 × 360 = 54 × total. So 9 × 360/54 = total. So 60 = total. For 1 pet: 48/360 × 60 = 8; 2 pets: 150/360 × 60 = 25; 3 pets: 84/360 × 60 = 14; 4 pets: 24/360 × 60 = 4. d) fx values: 0; 8;50;42;16 Mean = (0 + 8 + 50 + 42 + 16)/(9 + 8 + 25 + 14 + 4) = 116/60 = 1.9333. pets. **2** a) Negative correlation. b) He is describing positive correlation which is scatter diagram A.

Probability

10 minutes

Test your knowledge

1 Emike is playing a game with a spinner. The probabilities of getting each colour are shown in the table below.

Colour	RED	BLUE	GREEN	YELLOW
Probability	0.1	0.3		0.2

a Calculate the probability that Emike gets a green.

b What is the probability she gets a blue or a red?

c Emike wins the game if she gets a blue on the first spin and a yellow on the second. What is the probability that she wins?

d If you were to spin the spinner 500 times, how many times would you expect to get a red?

2 Cath catches the bus in the morning and in the evening. The probability the bus is late in the morning is 0.4; but in the evening is 0.3.

a Complete the following tree diagram.

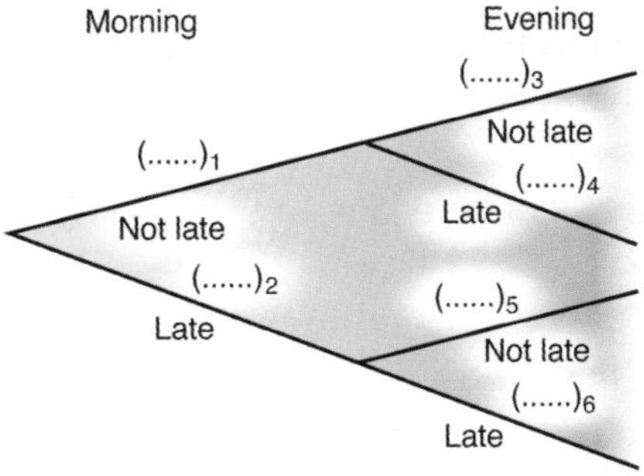

b Calculate the probability that neither bus is late.

Answers

1 a) 0.4. b) 0.4. c) 0.06. d) 50. 2 a) $(0.6)_1$, $(0.4)_2$, $(0.7)_3$, $(0.3)_4$, $(0.7)_5$, $(0.3)_6$. b) 0.42.

If you got them all right, skip to page 73

Probability

30 minutes

Improve your knowledge

Probability is about the chance that things happen. It is measured on a scale from 0 to 1. All the probabilities together must add up to 1.

1 Rules

There are 4 rules to help you work out probabilities. They are the **not** rule, the **and** rule, the **or** rule and the **number** rule.

a The **not** rule says that: probability of something **not** happening = 1 – probability the thing happens.

Example The probability it is sunny is $\frac{1}{10}$. What is the probability that it is not sunny?

Probability it is not sunny = 1 – probability it is sunny
= $1 - \frac{1}{10} = \frac{9}{10}$

b The **and** rule says that: you **multiply** the probabilities when you see the words '**both**' or '**and**'.

Example The probability that Bharat is late for a lesson is $\frac{1}{5}$. The probability that Riffat is late for a lesson is $\frac{3}{10}$. What is the probability that both Bharat and Riffat are late for a lesson?

Probability Bharat **and** Riffat are late = $\frac{1}{5} \times \frac{3}{10} = \frac{3}{50}$

AND means MULTIPLY.

c The **or** rule says that: you **add** the probabilities when you see the words '**either**' or '**or**'.

Example The probability Sarah travels to work by bus is 0.45. The probability she travels to work by train is 0.30. What is the probability that Sarah goes to work either by bus or by train?

Probability she travels by bus **or** train = 0.45 + 0.30
= 0.75.

OR means ADD.

d The **number** rule says that: **number** of times we expect something to happen = probability × total number of tries.

Example I throw a fair dice 300 times. Approximately how many threes do I expect to get?

The probability of a 3 is $\frac{1}{6}$.

So the **number** of 3s I expect = $\frac{1}{6} \times 300 = 50$

Why is the word approximately used? Answer is below.

2 Tree diagrams

Tree diagrams are a way of listing all the possibilities.

Example Natalie and Petrina play two games of darts. In the past, the probability that Natalie wins a game of darts is $\frac{3}{10}$.

Draw a tree diagram to represent the information.

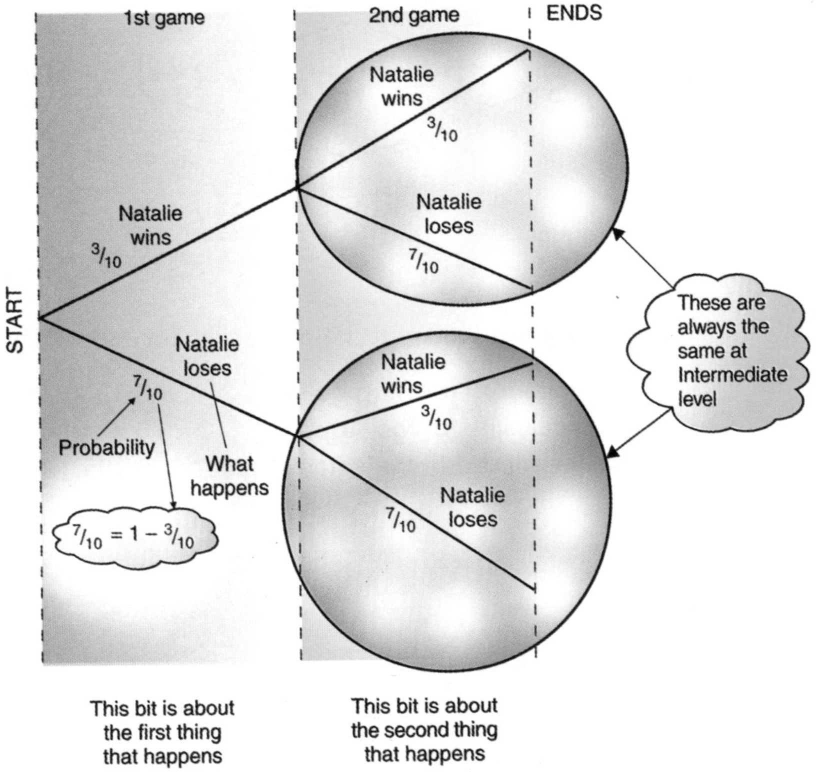

Answer

Because there do not HAVE to be 50 threes – there could be more or less.

To get each of the possibilities, you have to follow a 'path' through the tree diagram. A path must go from the start to one of the ends. In this diagram, since there are four ends, there are four paths.

To find the probability for a path, you must 'times out' the probabilities on it. So the probability for the top path ('Natalie wins, Natalie wins') is $\frac{3}{10} \times \frac{3}{10} = \frac{9}{100}$
.

What are the probabilities for the other paths?

Example When Jane goes to the art shop in her car, she needs to drive through two sets of traffic lights. The probability she has to stop at the first set is 0.3. The probability she has to stop at the second set is 0.65.

a Draw a tree diagram to show the different possibilities and calculate the probability that Jane has to stop at exactly one set of lights.

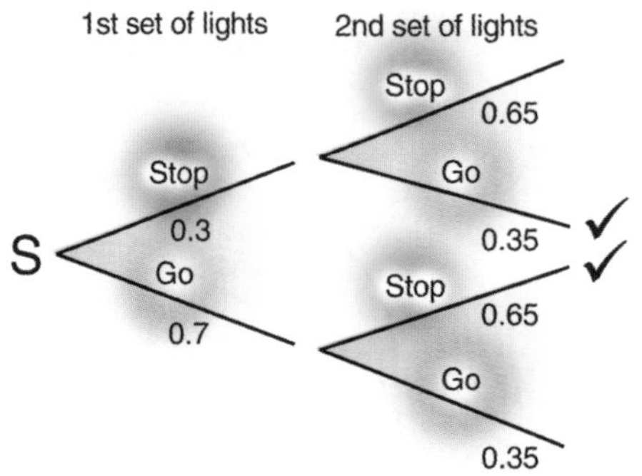

First we must decide which paths we want.
We want the path 'Stop Go' and the path 'Go Stop' shown with ✓.

We must first work out the probability for each of these paths:

The probability for 'stop go' is	$0.3 \times 0.35 = 0.105$.
The probability for 'go stop' is	$0.7 \times 0.65 = 0.455$.
Now we must add the different paths	$0.105 + 0.455 = 0.560$.

Remember – ADD DIFFERENT PATHS.

Answer

Second path: 'Natalie wins, Natalie loses' probability = $^{21}/_{100}$.
Third path: 'Natalie loses, Natalie wins' probability = $^{21}/_{100}$.
Fourth path: 'Natalie loses, Natalie loses' probability = $^{49}/_{100}$.

Probability

Use your knowledge

10 minutes

1 Simon and Hirminder play a game of squash and a game of chess. From previous experience, they know that the probability Simon wins squash is 0.55 and the probability Simon wins chess is 0.3.

a Complete the following tree diagram. Hints 1/2

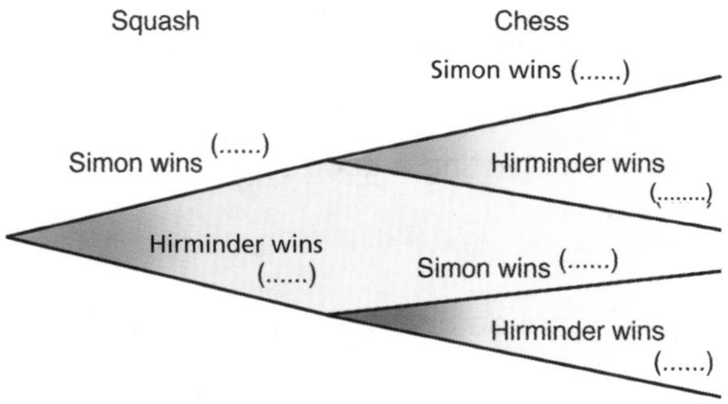

b What is the probability Hirminder wins at least 1 game? Hints 3/4

2 Nirmesh and Daksha are playing a game with dice. A 'good score' is a 5 or a 6.

a What is the probability of not getting a good score? Hints 5/6

b To play the game, two dice are thrown at once. The first person to get a 'good score' on both dice at once will win. Daksha goes first. What is the probability that she wins on her first go? Hints 7/8

c Nirmesh only has a go if Daksha does not win on her first go. What is the probability that he wins on his first go? Hints 9/10

Probability

Hints

1 Remember Hirminder wins if Simon does not win.

2 Use the '**not**' rule.

3 'At least 1' means either exactly one game or both games. Which paths do you need?

4 Remember to multiply to find the probability for each path, and add the different paths up.

5 What is the probability of getting a good score?

6 Use the **not** rule.

7 She must get a good score on both the 1st dice **and** the 2nd dice.

8 Use the **and** rule.

9 We need Daksha not to win on her first go **and** Nirmesh to win on his first go.

10 What is the probability Daksha does NOT win on her first go?

Answers

1 a) Reading left to right and downwards: 0.55; 0.3; 0.7; 0.45; 0.3; 0.7. b) $(0.55 \times 0.7) + (0.45 \times 0.3) + (0.45 \times 0.7) = 0.835$. 2) a) $1 - P(\text{good score}) = 1 - \frac{1}{3} = \frac{2}{3}$. b) Good score on first and second dice: $\frac{1}{3} \times \frac{1}{3} = \frac{1}{9}$. c) Need: Daksha does not win on her first go and Nirmesh wins on his first go. P(Daksha does not win on her first go) $= 1 - \frac{1}{9} = \frac{8}{9}$. P(Nirmesh wins on his first go $= \frac{1}{9}$). So probability we need $= \frac{8}{9} \times \frac{1}{9} = \frac{8}{81}$.

Ratios and similar triangles

10 minutes

Test your knowledge

1 ***a*** Jenny and Darsha are sharing a 1000 cm^3 bottle of Coke. They share it in the ratio 3 : 2. How much Coke does each of them get?

b Satinder and Matt are comparing their pocket money. They find that the amounts they get are in the ratio 3 : 5. If Matt gets £7.50, how much does Satinder get?

2 Explain why triangles ABC and ADE are similar.

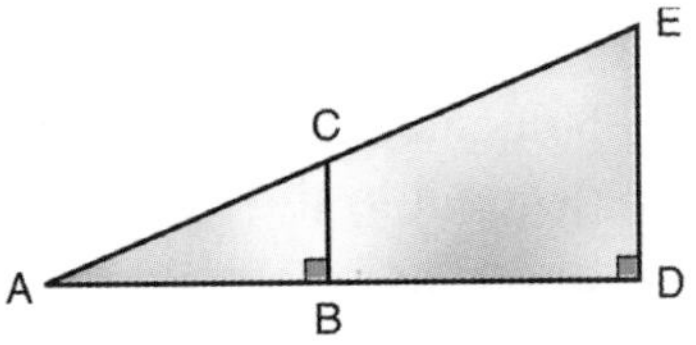

3 Triangles ABC and CDE are similar. Find X.

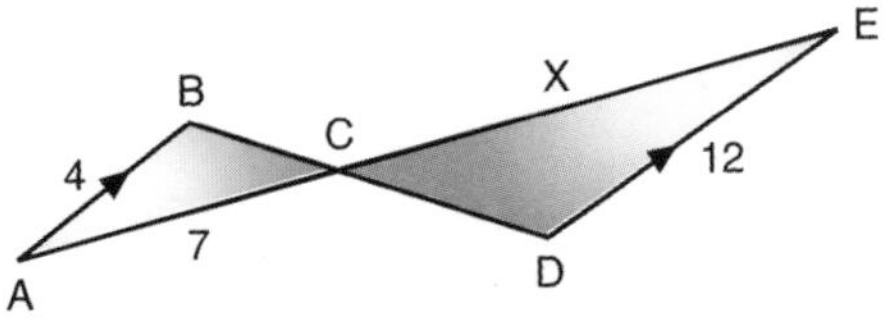

Answers

1 a) 600 cm^3 Jenny, 400 cm^3 Darsha b) £4.50. **2** Angles CAB and DAE are the same (they are the same angle) and angles ABC and ADE are both right angles. **3** 21.

If you got them all right, skip to page 79

Ratios and similar triangles

30 minutes

Improve your knowledge

1 Ratios

Ratios are about how amounts relate to each other – for example, if you eat twice as much cake as your sister, the ratio of the amounts you eat is 2 : 1. Ratios always look like two whole numbers with a colon (:) in between them. There are two types of ratio question.

Type 1

You are given a total amount and are asked to divide it in a particular ratio.

Example Sinneata, Laura and Nasser are given £50. They have agreed to divide it in the ratio 5 : 3 : 2. How much money does each of them get?

Step 1 Add up the 'ratio numbers' to get a 'total ratio number'.

The ratio numbers are 5, 3 and 2.
5 + 3 + 2 = 10, so the total ratio number is 10.

Step 2 Write down as a fraction

$$\frac{\left(\text{ratio number for the person you want}\right)}{\text{total ratio number}}$$

We'll do Sinneata first.
Write down $\frac{5}{10}$

Step 3 Multiply the fraction you have just written down by the total amount you were given in the question.
This is the answer!

We multiply $\frac{5}{10}$ by £50.

Don't worry about cancelling!

$\frac{5}{10} \times £50 = £25$.

So Sinneata's share is £25.

Step 4 Repeat to find other people's amounts.

We'll do Laura next.
We need $\frac{3}{10} \times £50 = £15$.
Now for Nasser.
$\frac{2}{10} \times £50 = £10$.

Step 5 Check the amounts add up! $£25 + £15 + £10 = £50$. ✓

Type 2

You are given the amount one person gets and asked to find what another person gets.

We can use the 2 × 2 table talked about in 'Number work' to answer this type of question:

Example Iain and Jaspal share some cakes in the ration 4 : 3. Jaspal gets 12 cakes. How many does Iain get?

We know Jaspal gets the ratio number 3 which means 12 cakes. Iain gets the ratio number 4. We need to work out how many cakes Iain gets.

Step 1 Draw a 2 × 2 table

Step 2 Fill in

Step 3 Draw a cross

Step 4 Calculation

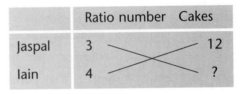

	Ratio number	Cakes
Jaspal	3	12
Iain	4	?

So answer = $\frac{12 \times 4}{3}$ = 16 cakes

Remember to check which type of ratio question it is before you start!

2 Similar triangles

Similar triangles aren't just triangles that look the same! They don't have to be identical – one can be bigger than the other – but …

Triangles are similar if 2 angles in one match 2 angles in the other.

Questions will ask you to do one of two things – either prove two triangles are similar or use the fact that triangles are similar to work out sides or angles.

Proving triangles are similar

To do this, you must find 2 angles in one that are the same as 2 angles in the other. You may have to use other things you know about angles – e.g. Z angles – to do this.

Example Show triangles ABC and CDE are similar.

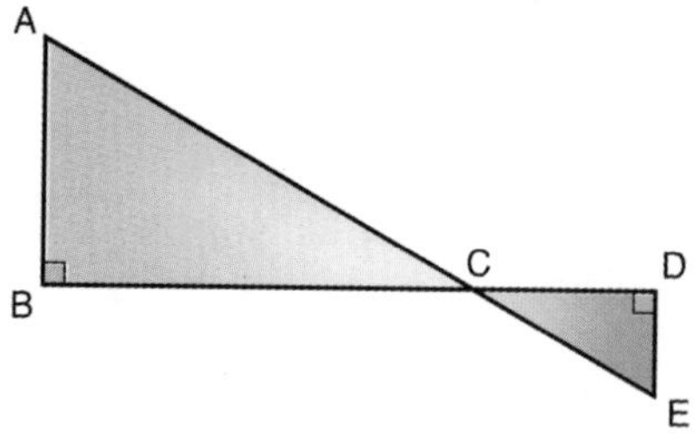

Angle CDE = Angle ABC (they are both right angles).
Angle ACB = Angle DCE (opposite angles).
So the triangles are similar.

3 *Using similar triangles*

Example ΔABC is similar to ΔEBD. Find the length x.

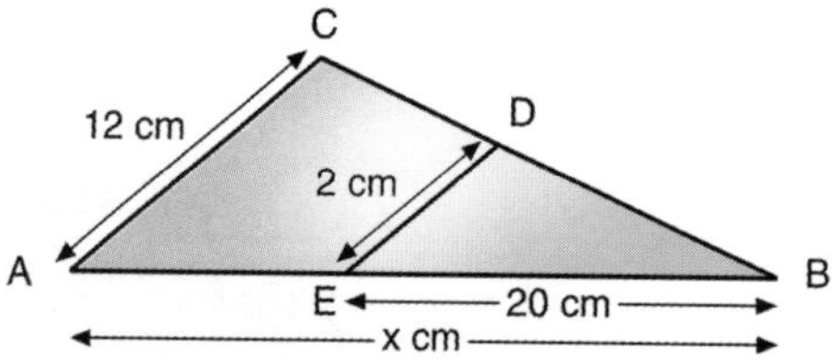

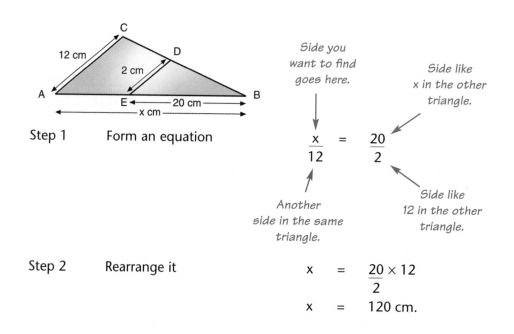

Step 1 Form an equation

Side you want to find goes here.

Side like x in the other triangle.

$$\frac{x}{12} = \frac{20}{2}$$

Another side in the same triangle.

Side like 12 in the other triangle.

Step 2 Rearrange it

$$x = \frac{20}{2} \times 12$$

$$x = 120 \text{ cm.}$$

✓ **Now learn how to use your knowledge**

Ratios and similar triangles

10 minutes

Use your knowledge

1 When Heena makes biscuits, she shares them with Mal.

a On Monday, Heena has 3 biscuits and Mal has 9 biscuits. Explain why this is a 1 : 3 ratio. **Hints 1/2**

Heena and Mal always share their biscuits in this same ratio.

b On Tuesday, Mal gets 12 biscuits. How many biscuits were there altogether on that day? **Hints 3/4**

c On Wednesday, Mal thinks he has had 7 biscuits. Heena tells him that this is impossible. **Hints 5/6**

Why is it impossible?

2 Heather and Ali have made a see-saw. But they haven't managed to balance it in the middle!

When they are on it, this is what happens.

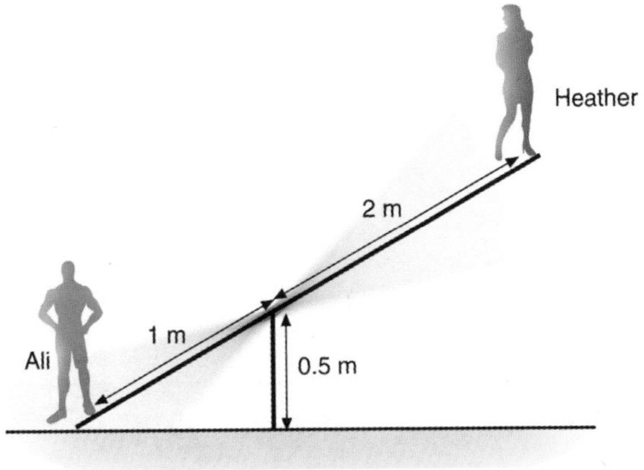

How high off the ground is Heather? **Hints 7/8/9/10**

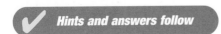

Hints and answers follow

Ratios and similar triangles

Hints

1 Write down a ratio using the numbers of biscuits you are given.

2 Simplify your ratio by cancelling.

3 Work out how many biscuits Heena had on Tuesday (remember you know the ratio).

4 Remember they eat all the biscuits between them.

5 Work out how many biscuits Heena would have had if Mal was right.

6 Can you make a fraction of a biscuit?

7 Look for a pair of similar triangles.

8 Make a big triangle in the diagram by drawing a vertical line between the top of the see-saw, where Heather is, and the ground directly below her.

9 What side of the big triangle do you need to know?

10 You can work out another side of the big triangle by adding two lengths together. What are they?

Answers

1 a) It is a 3 : 9 ratio. Cancelling threes gives 1 : 3. b) If Mal gets 12, then the number Heena gets is 12 × ⅓, so Heena gets 4. So there were 16 altogether. c) If Mal was right, Heena would get ⅓ × 7 = 2⅓ biscuits. But this would mean Heena has to have made fractions of a biscuit – which can't happen.

2 The longest side of the big triangle is 1 m + 2 m = 3 m. The big triangle and the smaller triangle are similar. So, using the similar triangle formula:

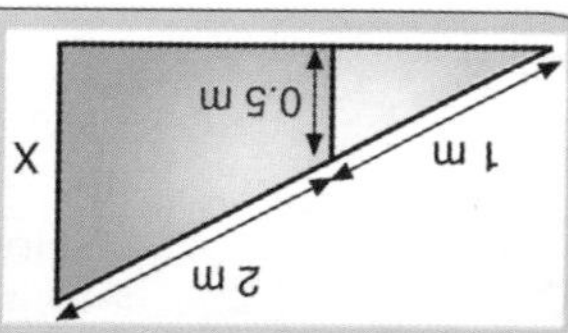

$$\frac{X}{3} = \frac{0.5}{1}$$

So X = 1.5 m.

Algebra 2

10 minutes

Test your knowledge

1 **a** Expand the brackets $2x(3x - 2y)$.

b Expand the brackets $(x + 5)(x - 2)$.

2 Factorise fully $15x^2 - 10xy$.

3 **a** Solve the simultaneous equations

$2x + 3y = 13$
$3x - y = 3$.

b Solve the simultaneous equations

$2x + 5y = 8$
$7x + 3y = -1$.

Answers

1 a) $6x^2 - 4xy$. b) $x^2 + 3x - 10$. **2** $5x(3x - 2y)$.
3 a) $x = 2, y = 3$ b) $x = -1, y = 2$.

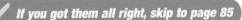

If you got them all right, skip to page 85

Algebra 2

Improve your knowledge

30 minutes

In this part of algebra, you will be looking at expanding brackets, factorising and simultaneous equations.

1 Expanding brackets

There are two sorts of questions on this – one involving one bracket and a numbe[r] or letter, and the other kind involving two brackets. Here's the first kind

Example 1 $2(p - q) = 2p - 2q$

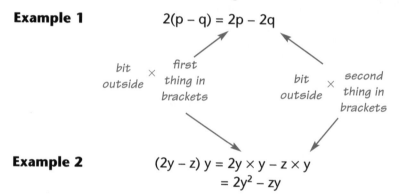

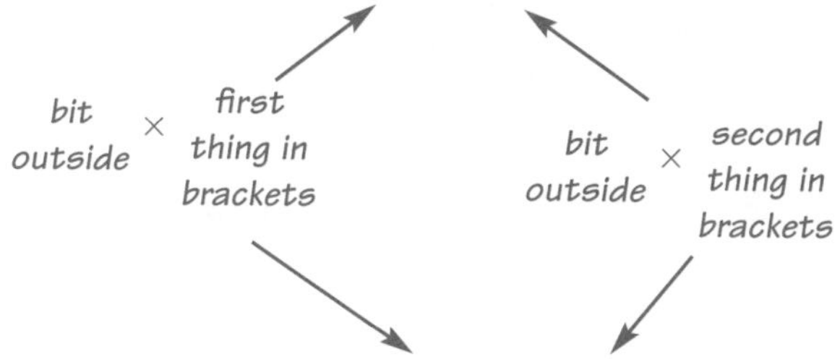

Example 2 $(2y - z)\,y = 2y \times y - z \times y$
$= 2y^2 - zy$

The other kind will look something like this:

$$(x + 2)(x + 3)$$

This looks a bit scary. However, we use the magic word **FOIL**.

FOIL stands for First, Outside, Inside and Last.
That means you times together the first term in each bracket,
outside terms,
inside terms,
last term in each bracket.

Let's try it ...

$(x - 1)(x + 4)$

First	$x \times x = x^2$,
Outside	$x \times 4 = 4x$,
Inside	$-1 \times x = -x$,
Last	$-1 \times 4 = -4$.

So we get $x^2 + 4x - x - 4 = x^2 + 3x - 4$.

2 Factorising

Factorising is just putting things back into brackets.

To do this, you need to look at the letters and numbers in two terms.

Letters	look for letters that are in both terms.
Numbers	look for numbers that divide into the numbers in both terms.

These are the things that will go **outside** the bracket.

Now we need to work out what goes inside the brackets.
First cancel out the letters that go outside.
Then divide by the number that goes outside.

Example 1 Factorise 6xy – 9y.

Letters	there is a y in both terms.
Number	3 goes into 6 and 9.

This tells us that **3y** goes outside the brackets.
So 6xy – 9y = 3y (something).

We need to work out what goes inside.

Letters first	cancel out the y:	$6x\cancel{y} - 9\cancel{y}$
Now divide by the 3	2x – 3.	

So 2x – 3 goes inside.

So we've got 3y(2x – 3).

Example 2 Factorise $2y^2 - 8y$.

We've got y in both terms and 2 goes into both numbers.
So outside the brackets we've got 2y.

Letters first	cancel out the y:	$2y^{\cancel{2}} - 8\cancel{y}$
Now divide by 2		y – 4

So we end up with 2y(y – 4).

When you cancel a y from y^2, you get left with y.

3 Simultaneous equations

Simultaneous equations are when you've got two equations with two letters in them. They will look something like this

$$2y + 3z = 5$$
$$4z - y = 3.$$

We are aiming to get rid of one of the letters by combining the equations.

There are steps we have to go through.

Step 1	Choose which letter we want to get rid of.	We'll choose to get rid of z (it could have been y).
Step 2	Write the equations so that the letters are in the same order.	$2y + 3z = 5$ $-y + 4z = 3$
Step 3	Look at the numbers in front of the letter you want to get rid of.	We want to get rid of z, 3 is in front of z in 1st equation, 4 is in front of z in 2nd equation.
Step 4	Times each equation by the number in front in the **other** equation.	We need to times the first equation by 4 and the second equation by 3. This gives us: $8y + 12z = 20$ $-3y + 12z = 9$.
Step 5	Look at the **signs** in front of the letter you want to get rid of. If they're the **same**, subtract the equations. If they're **different**, add the equations.	The signs are the **same**, so subtract. $8y - -3y + 12z - 12z = 20 - 9$ $11y = 11$ So $y = 1$.
Step 6	Put the value you have found back into one of the equations you started with.	Put $y = 1$ into $2y + 3z = 5$ Gives: $2 + 3z = 5$ So $3z = 3$ $z = 1$.

Look – only one letter left now.

Now learn how to use your knowledge

Algebra 2

10 minutes

Use your knowledge

1 Kaljit likes setting her friends puzzles. She tells Jinnie:

'If I buy 10 Fruit Chews and 7 Rhubarb and Custards, it costs me 41p.

If I buy 7 Fruit Chews and 10 Rhubarb and Custards, it costs me 44p.

How much would it cost me to buy 12 Rhubarb and Custards and 8 Fruit Chews?'

Jinnie decides to use F for the price of a Fruit Chew and R for the price of a Rhubarb and Custard. She then writes down two equations.

a What were Jinnie's equations? Hints 1/2

b Solve the equations to find the cost of each kind of sweet. Hints 3/4

c Hence answer Kaljit's question. Hint 5

2 Kaljit tries a different puzzle on another friend, Puja. She says:

'My garden is 2x – 5 metres long and x – 3 metres wide. If I walk all the way around the outside of it, I know I walk 44 metres. What is the length of the garden?

Puja starts by working out the perimeter of the garden in terms of x.

a Explain why she does this, and write down what she gets. Hints 6/7/8

b Puja then writes down an equation. What is it? Hint 9

c Solve this equation to find the value of x, and hence the length of the garden. Hints 10/11

d Puja also wants to work out the area. Write down an expression **in terms of x** for the area. Hints 12/13

e Use the value of x you found earlier to work out what the area is. Hint 14

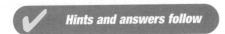

Hints and answers follow

Algebra 2

Hints

1. Work out how much the Fruit Chews and Rhubarb and Custards cost separately then add them together.
2. How much do 10 Fruit Chews cost in terms of F?
3. What sort of equations are they?
4. Try multiplying one equation by 7 and the other by 10, then subtracting them.
5. Use your answers for F and R.
6. What is the same as the perimeter in the question?
7. Remember that you find the perimeter by adding together all the sides.
8. Remember you need two lots of the length and two lots of the width.
9. The perimeter needs to be 44 metres.
10. First take the minus number over, then get rid of the number in front of x.
11. If x takes this value, how long is the garden?
12. Area = length × width.
13. Use FOIL.
14. Put your value of x into your formula for the area.

Answers

1 a) $10F + 7R = 41$, $7F + 10R = 44$. b) Multiply 1st equation by 7. $70F + 49R = 287$. Multiply 2nd equation by 10. $70F + 100R = 440$ (this could be other way round). Subtract 1st from 2nd: $51R = 153$, so $R = 153/51 = 3$. Put R=3 back into one of the original equations $10F + 21 = 41$. So $10F = 20$. So $F = 20/10 = 2$. c) We want the cost of 12R and $8F = 12 \times 3 + 8 \times 2 = 52$p. **2** a) Because she has been given some information about the perimeter – when Kaljit is walking around the outside, the distance she walks is the perimeter: $2x - 5 + 2x - 5 + x - 3 + x - 3 = 6x - 16$. b) $6x - 16 = 44$. c) $6x = 44 + 16$ so $6x = 60$ so $x = 10$ metres. Length is $2x - 5$. Putting in $x = 10$ gives 15 metres. d) Area = length × width = $(2x - 5) \times (x - 3) = 2x^2 - 6x - 5x + 15 = 2x^2 - 11x + 15$. e) $x = 10$. So area $2 \times 10^2 - 11 \times 10 + 15 = 105m^2$.

Mock exam

1 hour 30 minutes

1 The household expenditure for the Smith family is illustrated on the pie chart below.

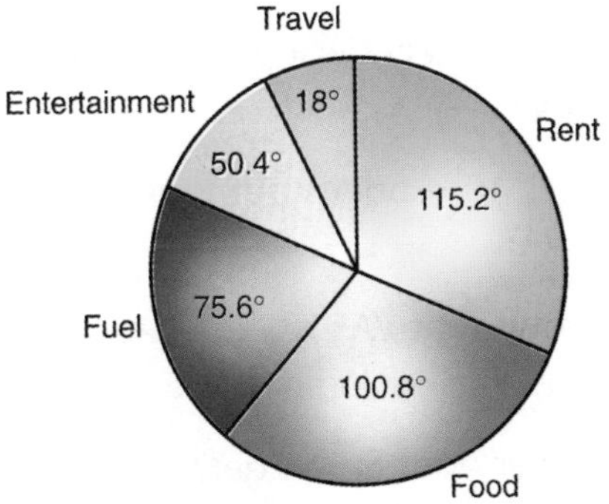

a Explain why this tells you that the family spends twice as much on food as on entertainment.

....

.... **(2)**

b What **percentage** of their income do the Smiths spend on **food and fuel put together**?

....

.... **(3)**

c The total amount the Smith family spend each week is £250. How much per week do they spend on food?

....

.... **(2)**

2 Imran has saved up some money to buy a Hi-Fi. He looks around the shops to try to find the best bargain.
He sees these two notices.

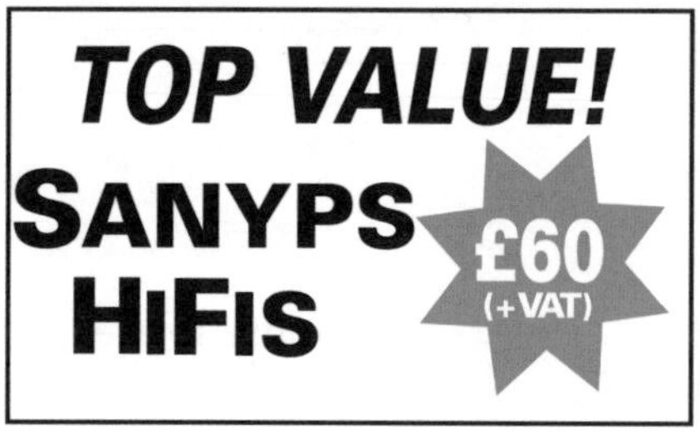

Imran immediately decides he will buy the Sanyps Hi-Fi because it is cheaper.

His brother Asim points out that he must be careful because just one of the prices includes VAT.

Imran knows that VAT is added at 17.5%, so he decides to work out the price of the Sanyps Hi-Fi including VAT.

What answer does Imran get?

....

.... **(3)**

3 The following instructions are used to work out a number sequence:

Step 1 Write down the number 11
Step 2 Subtract 3 from the number you have just written
Step 3 Write down the number obtained in step 2
Step 4 If your answer in step 2 is less than 2 then stop
Step 5 Go to step 2.

a Write down the numbers produced from the set of steps.

....

.... **(3)**

b Change one line of the set of steps so that it could be used to produce the first 6 odd numbers.

.... **(1)**

4 ABCDEFGH is a box. AB = CD = EF = GH = 4 cm. AE = BF = CG = DH = 2 cm. AD = BC = FG = EH = 3 cm.

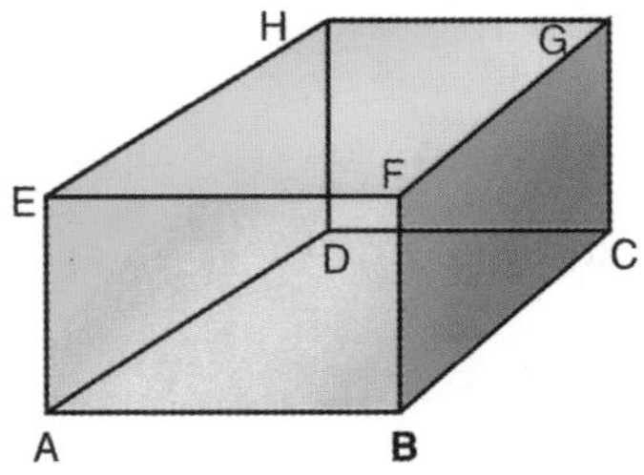

a Show that the length of AC is 5 cm

.... **(2)**

Use triangle ACG to:

b Find the length of AG

.... **(2)**

c Find angle GAC.

.... **(2)**

5 Sue's pocket money has gone up by 15%. She now gets £6.90 every week. How much money did she get each week **before** it went up?

....

.... **(3)**

6 Justine goes out on Friday and Saturday nights. She only goes out to one place each night.

The probability that she will go to the cinema on any one day is 0.45.

a What is the probability that Justine will not go to the cinema on Friday?

.... **(1)**

The probability that Justine goes to the disco on any one day is 0.2.

b What is the probability that Justine either goes to the cinema or goes to the disco on Friday?

.... **(2)**

c What is the probability that Justine goes to the disco on Friday and to the cinema on Saturday?

.... **(2)**

7 At 'Jeans R Us', the manager recorded the number of men's jeans of each size sold in a week:

Jeans size	26–28	30	32	34	36–40
No of pairs sold	4	6	8	7	5

a Calculate an estimate of the mean jeans size.

....

....

.... **(3)**

b Explain why this answer is only an estimate.

....

.... **(1)**

c State the median.

.... **(2)**

8 The table below shows the minimum and maximum amounts (in pounds) that five students had in their bank accounts over one year (a negative amount means the student had an overdraft).

Student	Minimum amount (£)	Maximum amount (£)
Mandeep	28	1300
Laura	–10	700
Richard	–200	250
Damien	600	2800
Sandhya	–50	750

a Which student had the largest overdraft?

.... **(1)**

b What is the difference between the maximum and minimum amounts in Sandhya's account?

.... **(1)**

9 Tanith has to pack tubes of sweets into boxes. Each box holds at most 28 tubes. She always uses the smallest possible number of boxes. One morning, Tanith has 267 tubes.

a Write down a calculation which Tanith should use to find the number of boxes.

.... **(2)**

b How many boxes should Tanith use?

.... **(1)**

10 In the diagram below AD is parallel to BC.

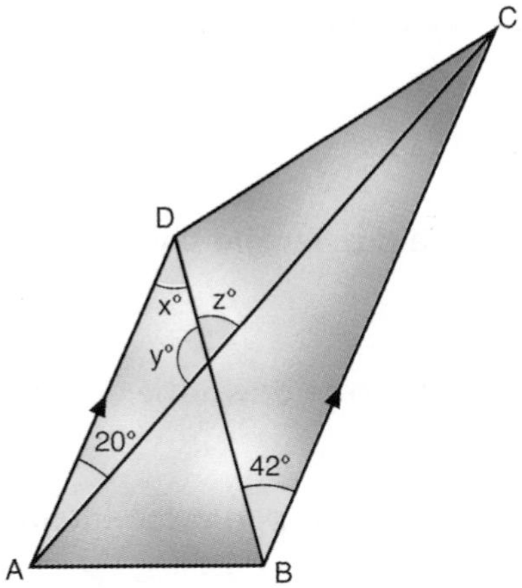

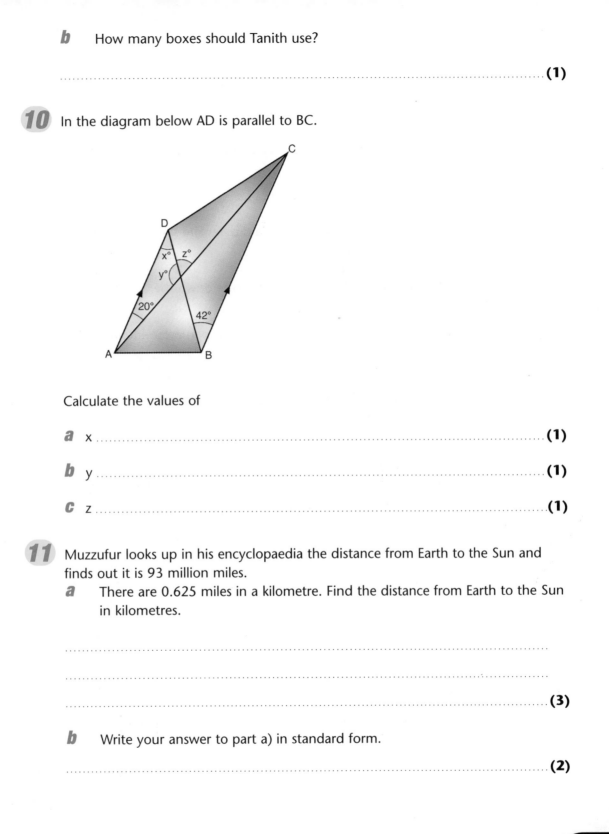

Calculate the values of

a x .... **(1)**

b y .... **(1)**

c z .... **(1)**

11 Muzzufur looks up in his encyclopaedia the distance from Earth to the Sun and finds out it is 93 million miles.

a There are 0.625 miles in a kilometre. Find the distance from Earth to the Sun in kilometres.

....

....

.... **(3)**

b Write your answer to part a) in standard form.

.... **(2)**

12 *a* Use your calculator to find the value of

$$\frac{6.22 \times \sqrt{3.9}}{32.3 + 1.9}$$

.... **(1)**

b Give your answer to a) correct to 1 decimal place.

.... **(1)**

13 Metale sets her friend Rizwan a puzzle. She says 'My lawn is a rectangle 2x metres long and x – 3 metres wide.
Its area is 22.88 m^2. What is x?'
The first thing Rizwan does is to write down the equation
$2x(x - 3) = 22.88$.

a Explain why he writes down this particular equation.

.... **(1)**

Rizwan then gets stuck! Metale gives him a hint: 'x is between 4 and 6'. He decides to try out some numbers.

First he tries $x = 5$ and gets the result 20 for $2x(x - 3)$.
Then he tries $x = 6$ and gets the result 36 for $2x(x - 3)$.

b By trying other values of x and showing your working clearly, find a solution of the equation correct to 1 decimal place.

....

....

....

.... **(3)**

14 Sita is learning physics. She wants to use the formula
$2as = v^2 - u^2$.

a She wants to know the value of a when $s = 2.8$, $v = 6.1$ and $u = 0.9$.
Find this value.

....

.... **(2)**

b She then needs to work out a value of u when she knows the value of the other letters.
Make u the subject of the formula.

....

....

....

....**(3)**

15 Factorise fully $6a^2b - 9ab^2$.

....

....**(2)**

16 **a** Complete the table below.

x	–3	–2	–1	0	1	2	3
$x^2 + 2$	11			2	3		

(4)

b Hence draw the graph of $y = x^2 + 2$ on the graph paper below.

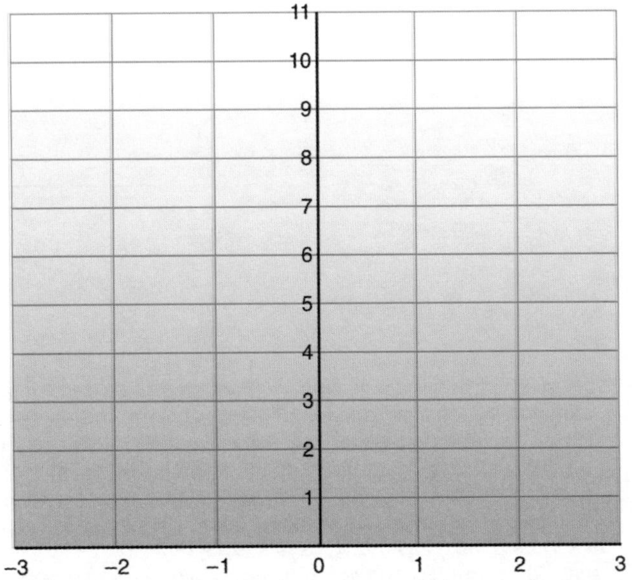

(2)

c On the same graph paper, draw the graph of the straight line $y = x + 4$. **(3)**

d Hence write down the two solutions of the simultaneous equations

$$y = x + 4$$
$$y = x^2 + 2$$

.... **(2)**

17 The cylinder shown has diameter 10 cm and height 30 cm. Work out its volume.

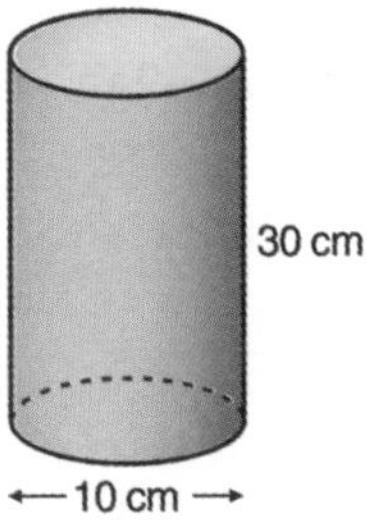

....

.... **(2)**

Answers

1 a) Because the angle for food is twice as big as the angle for entertainment. b) 49%. c) £70. **2** £70.50 **3** a) 11, 8, 5, 2, –1. b) Step 2 changes to 'Subtract 2 from the number you've just written'. **4** a) $\sqrt{(3^2 + 4^2)} = \sqrt{25} = 5$ cm. b) 5.385 cm. c) 21.8°. **5** £6. **6** a) 0.55. b) 0.65. c) 0.09. **7** a) 32.4. b) It is only an estimate because the exact number of each size is not known for sizes 26, 28, 36, 38, 40. c) 32. **8** a) Richard. b) £800. **9** a) 267 divided by 28. b) 9. **10** a) 42. b) 118. c) 62. **11** a) 148 800 000 km. b) 1.488×10^8 km. **12** a) 0.359 167. b) 0.4 **13** a) Because the area = length × width = 2x (x – 3), and area = 22.88. b) 5.2. **14** a) 6.5. b) $u = \sqrt{(v^2 - 2as)}$. **15** 3ab(2a – 3b). **16** a) Missing figures are 6,3,6,11.

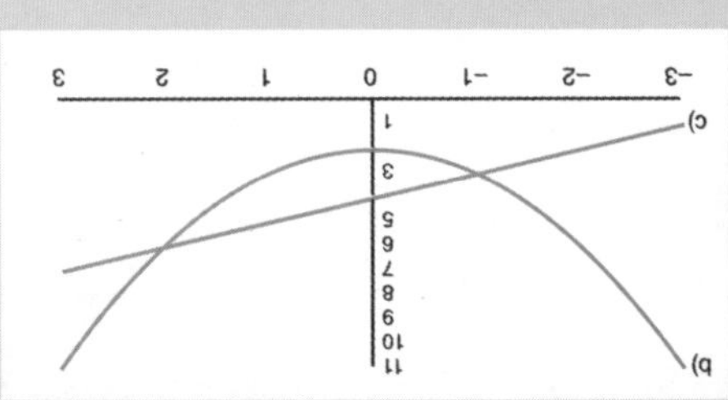

d) x = 2, y = 6 and x = –1, y = 3.

17 2356.1945.